Modern Operations in Pulping & Bleaching: A Paper Engineering Primer

Lucian A. Lucia, PhD
Brian N. Brogdon, PhD

Published by North Carolina State University Libraries

Distributed by the University of North Carolina Press

NC STATE University Libraries

This work was supported by grant funding from the NC State University Libraries' Alt-Textbook Project.

Suggested citation: Lucia, Lucian A. and Brian N. Brogdon. *Modern Operations in Pulping and Bleaching: A Paper Engineering Primer.* Raleigh: North Carolina State University Libraries, 2025.

DOI: https://doi.org/10.5149/9781469696461_Lucia

ISBN 978-1-4696-9645-4 (paperback)
ISBN 978-1-4696-9646-1 (open access PDF ebook)

For product safety concerns under the European Union's General Product Safety Regulation (EU GPSR), please contact gpsr@mare-nostrum.co.uk or write to the University of North Carolina Press and Mare Nostrum Group B.V., Mauritskade 21D, 1091 GC Amsterdam, The Netherlands.

This book is dedicated to the memory of Peter W. Hart, whose mentorship and inspiration helped shape the careers of the authors and countless others in the pulp and paper community. His legacy lives on through the spirit of learning and discovery embodied in these pages.

Preface

The pulp & paper industry continues to evolve and flourish, driven by advances in science, engineering, and sustainability in this age of the rapidly emerging biorefinery. Yet at its core, the industry fundamentally depends on a solid appreciation of the key operations that transform wood into pulp and ultimately into paper products. This book, *Modern Operations in the Pulping and Bleaching: A Paper Engineering Primer*, was written to provide students of Paper Science & Engineering (PSE) with a clear, practical foundation in these operations from wood handling through bleaching.

Our aim is to bridge theory with practice, offering not only the key concepts that govern industrial processes but also guidance on how these principles can be applied in real-world laboratory and production settings. By grounding students in the essential calculations and operations of the field, we hope to prepare the next generation of engineers to contribute meaningfully to the industry.

This work is also a tribute and culmination of a promise. We dedicate this book to the loving memory of Professor Peter W. Hart, a long-time friend, colleague, and mentor who has long been involved in our teaching and research and whose work inspired us to write this text. His wisdom, generosity, and unwavering encouragement inspired us immensely and helped shape the trajectory of our careers. His memory continues to guide us, and we hope that the spirit of curiosity, rigor, and mentorship that he embodied is reflected throughout these pages. In addition to Dr. Hart, we extend our deep-felt gratitude to Dr. Hasan Jameel who led and spearheaded for many years the capstone PSE course on which our text is inspired, and from whom we graciously received several key pulping figures.

We trust that this book will serve as a contemporary and invaluable resource for generations of students, educators, and practitioners, and it will help carry forward the legacy of learning and innovation that has long characterized the pulp & paper community at North Carolina State University.

Lucian A. Lucia
Brian N. Brogdon

3

Chapter 1: Solutions

I. Acids/Bases

Introduction to Acid/Base Chemistry & Equilibria

With respect to modern pulp & paper industrial chemistry, acid/base chemistry is a topic encompassing the fundamental behaviors of acids and bases in water. Understanding this topic is crucial for grasping the chemical behavior of wood in pulping liquors. First, we must understand the nature of a solution. A solution is a system consisting of a completely dissolved substance within a liquid, in this case, water. For example, if we take 10 g of NaCl and dissolve it in water, we will still have 10 g of NaCl, but it will have dissolved into 10 g of solvated Na^+ AND Cl^- ions.

Definitions and Concepts

Acids and Bases

An acid is a substance that increases the concentration of hydrogen ions (H^+) in aqueous solution, while a base increases the concentration of hydroxide ions (OH^-) falls under the designation of Arrhenius.

Arrhenius Acids and Bases

$$HCl \longrightarrow H^+ + Cl^-$$

Acid - forms H^+ in water

$$NaOH \longrightarrow Na^+ + {}^-OH$$

Base - forms ^-OH in water

Figure 1.1. A chemical description of the nature of acids and bases according to Arrhenius.

More specifically, Brønsted-Lowry designates an acid as a proton donor, and a base as a proton acceptor.

Brønsted-Lowry Acids and Bases

Acid	Base		

$$HCl + NaHCO_3 \longrightarrow NaCl + CO_2 + H_2O$$

$$HBr + NH_3 \longrightarrow NH_4Br$$

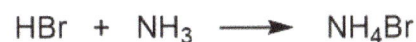

proton donor   proton acceptor

Figure 1.2. A more specific chemical description of acids and bases according to Brønsted-Lowry.

Conjugate Acids and Bases

Conjugate acids and bases are related pairs of compounds that differ by the presence or absence of a proton (H^+). When an acid donates a proton, it forms its conjugate base, which is the species that remains after its proton is lost. Conversely, when a base accepts a proton, it forms its conjugate acid, which is the species that results from its gain of the proton. For example, in the reaction of acetic acid (CH_3COOH) with water, acetic acid donates a proton to water, forming the acetate ion (CH_3COO^-) as its conjugate base and the hydronium ion (H_3O^+) as the conjugate acid of water (please see Homework Problem 1.c).

Homework

1. For each of the following reactions, identify the Brønsted-Lowry acid and base:
   a) $HCl + H_2O \rightarrow H_3O^+ + Cl^-$
   b) $NH_3 + H_2O \rightarrow NH_4^+ + OH^-$
   c) $CH_3COOH + H_2O \rightarrow CH_3COO^- + H_3O$

2. Write the conjugate acid for each of the following bases:
   a) OH^-
   b) NH_3
   c) HCO_3^-

3. Write the conjugate base for each of the following acids:
   a) H_2SO_4
   b) H_2O
   c) H_3PO_4

4. Predict the products of the following Brønsted-Lowry acid-base reactions and identify the conjugate acid-base pairs:
   a) $HNO_3 + H_2O \rightarrow$
   b) $NH_4^+ + OH^- \rightarrow$
   c) $H_2CO_3 + H_2O \rightarrow$

pH and pOH

Solution pH is a measure of its acidity (protons) or alkalinity (hydroxides) defined as the negative logarithm of the hydrogen ion concentration:

$$pH = -\log [H^+]$$

Eq. 1

Similarly, pOH is defined as the negative logarithm of the hydroxide ion concentration:

$$pOH = -\log [OH^-]$$

Eq. 2

The relationship between pH and pOH in water at 25°C is:

$$pH + pOH = 14$$

Eq. 3

Strong and Weak Acids/Bases

Strong Acids/Bases are substances which go into water and completely dissociate (acids yield all their hydrogens, while bases yield all their hydroxides). Acid and base examples include hydrochloric acid (HCl) and sodium hydroxide (NaOH), respectively. On the other hand, Weak Acids/Bases are substances which also go into solution, but only partially dissociate in water. Acid and base examples include acetic acid (CH_3COOH) and ammonia (NH_3), respectively.

Acid-Base Equilibria

In aqueous solutions, acids and bases establish an equilibrium between their dissociated (ionized) and undissociated (molecular) forms. This equilibrium can be described by an equilibrium constant.

Equilibrium Constants

For a weak acid HA dissociating a proton in water, the K_a (Acid Dissociation Constant) can be calculated as follows:

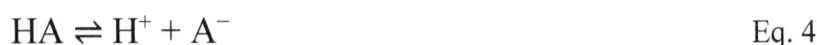

$$HA \rightleftharpoons H^+ + A^-$$

Eq. 4

The equilibrium constant expression is:

$$K_a = \frac{[H^+][A^-]}{[HA]}$$

Eq. 5

For a weak base B accepting a proton in water, the K_b (Base Dissociation Constant) can be calculated as follows:

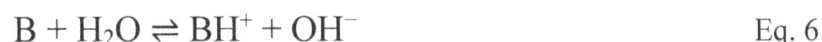

$$B + H_2O \rightleftharpoons BH^+ + OH^-$$

Eq. 6

7

The equilibrium constant expression is:

$$K_b = \frac{[BH^+][OH^-]}{[B]}$$

Eq. 7

Henderson-Hasselbach Equation

The Henderson-Hasselbalch equation is a useful formula for calculating pH of a buffer solution. It relates the pH of a solution to the pK_a (the negative logarithm of the weak acid dissociation constant, K_a) and the ratio of the concentrations of the conjugate base and the weak acid. The equation is expressed as:

$$pH = pK_a + \log\frac{[A^-]}{[HA]}$$

Eq. 8

In this equation, $[A^-]$ represents the concentration of the conjugate base, and $[HA]$ represents the concentration of the weak acid. The equation is derived from the acid dissociation constant expression for a weak acid in equilibrium previously shown on the previous page:

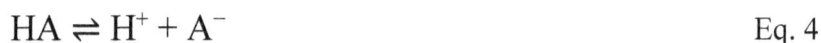

$$HA \rightleftharpoons H^+ + A^-$$

Eq. 4

By rearranging the equilibrium expression and taking the logarithm, the Henderson-Hasselbalch equation provides a straightforward way to calculate pH of a buffer solution, when we have the concentrations of the acid and its conjugate base. This equation is particularly useful in preparing buffer solutions and understanding how they resist changes in pH when extraneous small amounts of acid or base are added.

Le Chatelier's Principle

Le Chatelier's Principle states that if a dynamic equilibrium is disturbed by changing chemical conditions, the position of equilibrium moves to counteract the change. This principle helps predict how a system at equilibrium will respond to changes in concentration, temperature, or pressure. For example, if the concentration of a reactant is increased, the system will shift to consume the added reactant and produce more products. Similarly, if the temperature of an exothermic reaction is increased, the equilibrium will shift to favor the reactants, absorbing the added heat. This principle is essential for understanding and controlling chemical reactions in various industrial and laboratory processes.

Homework

1. Calculate the pH of a 0.1 M solution of acetic acid (CH_3COOH) given that the K_a of acetic acid is 1.8×10^{-5}.
2. A 0.05 M solution of a **weak acid** has a pH of 3.2. Calculate the K_a of the acid.

3. Calculate the pH of a buffer solution that contains 0.1 M acetic acid (CH_3COOH), the weak acid, and 0.1 M sodium acetate ($CH_3COO^-Na^+$), the conjugate base. The K_a of acetic acid is 1.8 x 10^{-5}.
4. Using the Henderson-Hasselbalch equation, calculate the pH of a buffer solution made by mixing 50 mL of 0.1 M NH_3 and 50 mL of 0.1 M NH_4Cl. The K_b of ammonia is 1.8 x 10^{-5}.
5. For the reaction: $NH_3 + H_2O \rightleftharpoons NH_4^+ + OH^-$ Given that the initial concentration of NH_3 is 0.1 M and the K_b of NH_3 is 1.8 x 10^{-5}, calculate the equilibrium concentrations of all species.
6. Explain how Le Chatelier's Principle applies to the following equilibrium when additional HCl is added: $CH_3COOH \rightleftharpoons CH_3COO^- + H^+$. Predict the direction of the shift in equilibrium and the resulting changes in concentrations.

Relationship Between K_a and K_b

For a conjugate acid-base pair, the product of their dissociation constants is equal to the ion-product constant for water (K_w):

$$K_a \times K_b = K_w; \text{ at } 25°C, K_w = 1.0 \times 10^{-14}$$ Eq. 9

pK_a and pK_b

The strength of an acid or base can also be expressed in terms of **pK_a** and **pK_b**, which are the negative logarithms of **K_a** and **K_b**, respectively:

$$\mathbf{pK_a = -logK_a}$$ Eq. 10

$$\mathbf{pK_b = -logK_b}$$ Eq. 11

Understanding these concepts and their associated equilibria is essential for predicting the behavior of acids and bases in various chemical contexts, from industrial processes to biological systems [1] [2] [3].

Molarity

Molarity, also known as molar concentration, is a measure of the concentration of a solute in a solution. It is defined as the number of moles of solute per liter of solution. The formula for calculating molarity is:

$$M = n/V$$ Eq. 12

where M is the molarity, n is the number of moles of the solute, and V is the volume of the solution in liters. For example, if you dissolve 1 mole of sodium hydroxide (NaOH) to make a 1 liter of water solution, the molarity of the solution is 1 M (1 mol/L). Another example is dissolving 0.5 moles of hydrochloric acid (HCl) to make 2 liters of water solution. The molarity of this solution is: M = 0.5 moles/2 liters = 0.25 M. Molarity is a crucial concept in chemistry because it allows chemists to quantify the concentration of solutions accurately, which is essential for reactions and experiment.

Equivalents & Normality

Normality (N) is a measure of concentration that expresses the number of equivalents of a solute per liter of solution. An equivalent is the amount of a substance that reacts with or supplies one mole of hydrogen ions (H^+) or hydroxide ions (OH^-) in a chemical reaction. This concept is particularly useful in acid-base chemistry and redox reactions, where the reactive capacity of a solute is more relevant than its molar concentration.

The formula for normality (N) is:

$$N = \text{Number of gram equivalents of solute/Volume of solution in liters.} \qquad \text{Eq. 13}$$

To calculate the number of gram equivalents, you can use:

$$\text{Number of gram equivalents} = \frac{\text{Weight of solute (grams)}}{\text{Equivalent weight of solute}} \qquad \text{Eq. 14}$$

The number of gram equivalents is a measure of the reactive capacity of a substance in a reaction. It is calculated by dividing the mass of the substance by its equivalent weight. The equivalent weight of a substance depends on the type of reaction it undergoes:

1. **For acids**: The equivalent weight is the molar mass divided by the number of hydrogen ions (H^+) the acid can donate.
2. **For bases**: The equivalent weight is the molar mass divided by the number of hydroxide ions (OH^-) the base can accept.
3. **For salts**: The equivalent weight is the molar mass divided by the total positive or negative charge of the ions produced when the salt dissociates.
4. **For redox reactions**: The equivalent weight is the molar mass divided by the number of electrons transferred in the reaction.

Examples: the case of sulfuric acid (H_2SO_4), which can donate two hydrogen ions, the equivalent weight is the molar mass of H_2SO_4 (98 g/mol) divided by 2, giving an equivalent weight of 49 g/equiv. Additionally, in a 1 M solution of sulfuric acid (H_2SO_4), the normality is 2 N because each molecule of H_2SO_4 can donate two hydrogen ions. Similarly, in a 1 M solution of sodium hydroxide (NaOH), the normality is 1 N because each molecule of NaOH can donate one hydroxide ion. Thus, the relationship between molarity and normality is:

$$M \times \text{no. of equivalents} = N \qquad \text{Eq. 15}$$

Homework

1. Calculate the normality of a solution containing 0.5 moles of sulfuric acid (H_2SO_4) in 1 liter of solution. (Hint: H_2SO_4 can donate 2 hydrogen ions per molecule.)
2. How many equivalents of hydrochloric acid (HCl) are present in 250 mL of a 2 N solution?
3. Calculate the normality of a solution prepared by dissolving 0.1 moles of sodium hydroxide (NaOH) in 500 mL of water.
4. A solution of phosphoric acid (H_3PO_4) has a molarity of 0.5 M. Calculate its normality, considering that H_3PO_4 can donate 3 hydrogen ions per molecule.

Weak Base (Na_2S) Equilibrium (First)

The following equilibrium is key to understanding where and how hydroxides are distributed in pulping systems:

$$\text{Eq. 16}$$

Please note that K_b is an equivalent equilibrium expression for the above equilibrium because the above equation expresses a base dissociation, which in this case is a weak base. In the above case, the $K_{eq} = 10$. However, as you can see, this weak base is "polyprotic", i.e., it can dissociate more than once. How? In this case, the hydrosulfide anion (HS^-) can also dissociate to another hydroxide. To determine how to capture the quantity of hydroxide, we will need the equilibrium constant (K_{eq} or K_b) AND the pH (which you may have surmised from the previous in this chapter).

How do we measure the quantities of hydroxide that will help us to acquire pH? We can titrate using a known concentration of strong acid (protons) given a base-indicator (HI + $^-$OH → I$^-$ [colored] + H_2O) monitored 100% efficient neutralization reaction (H^+ + $^-$OH → H_2O).

What is the actual equilibrium equation? We can deduce it from a comparison of the equilibrium species (ignoring spectators and solvent which do not affect the equilibrium):

$$K_{eq} = \frac{[HS^-][OH^-]}{[Na_2S]} \qquad \text{Eq. 17}$$

If $K_{eq} = 10$, then the ratio, $\frac{[HS^-][OH^-]}{[Na_2S]} = 10$, which means that there is 10 times more product than the weak acid. If we wish to isolate pH/pOH (see Eqs. 1, 2, and 3), we will need to isolate the hydroxide concentration whose pOH is a logarithmic (log base 10) expression (see Eq. 2). Therefore, we can take the log of both sides which will give (using the rules of logarithms):

$$\log K_{eq} = \log[HS^-] + \log[OH^-] - \log[Na_2S] \qquad \text{Eq. 18}$$

There are two quantities in this equation in their negative log expressions will provide us with pK_{eq}/pOH values. Therefore, if we multiply both sides of the equation by a -1, we obtain:

$$-\log K_{eq} = -\log[HS^-] - \log[OH^-] + \log[Na_2S] \qquad \text{Eq. 19}$$

This equation can be further reduced to:

$$pK_{eq} = -\log[HS^-] + pOH + \log[Na_2S] \qquad \text{Eq. 20}$$

According to the logarithmic rules (see earlier), we can further reduce this expression to:

$$pK_{eq} = pOH + \log\frac{[Na_2S]}{[HS^-]} \qquad \text{Eq. 21}$$

Using Eq. 3, we can rearrange the above equation to the following:

$$pK_{eq} = (14 - pH) + \log\frac{[Na_2S]}{[HS^-]} \qquad \text{Eq. 22}$$

In which if we know the pK_{eq} and the pH, we can obtain the concentration of hydroxide (hint: it is the same as [HS⁻], see Eq. 16). Because K_{eq} =10, pK_{eq} = -$\log K_{eq}$, which when solved gives -1. Therefore,

$$pK_{eq}(or - 1) = (14 - pH) + \log\frac{[Na_2S]}{[HS^-]} \qquad \text{Eq. 23}$$

$$-1 - 14 + pH = \log\frac{[Na_2S]}{[HS^-]} \qquad \text{Eq. 24}$$

$$-15 + pH = \log\frac{[Na_2S]}{[HS^-]} \qquad \text{Eq. 25}$$

Therefore, if pH is known, we can determine the distribution of hydroxide.

Homework

1. Calculate the percentages of hydroxide produced for the weak base dissociation of sodium sulfide at the following pH: 14, 13, 12, 10, and 9.
2. What is the trend in hydroxide as pH becomes more acidic?
3. Calculate the percentage of product formed or reactant remaining with the understanding percentage is calculated by the following formula:
   Y - 1/Y x 100, where Y = ratio as an order of 10; e.g., if Y = 100 (two orders of ten), then the percentage of product or reactant formed = (100 -1)/100 x 100 = 99%
4. Can you provide an explanation for the trend? Hint: Invoke LeChatelier's Principle.

Weak Base (Na₂S) Equilibrium (Second Constant)

The weak base, sodium sulfide, in pulping liquors is polyprotic, meaning it has more than simple activity. It can give dissociate into more than just one hydroxide. Take note of its second equilibrium:

Eq. 26

This second equilibrium is quite different than the first equilibrium. It is MUCH weaker, i.e., 8 orders of magnitude, in fact. The $K_{eq} = 10^{-7}$ which means as opposed to the first equilibrium, the reaction favors the reactant side by a far margin!

The equilibrium expression is:

$$K_{eq} = \frac{[H_2S][OH^-]}{[HS^-]}$$

Eq. 27

Which can be rearranged to provide the K_{eq} and the pH quantities needed as shown previously in Eqs. 18 to 25, which ultimately yield the expression:

$$pK_{eq} \text{ of second dissociation} - 14 + pH = \log\frac{[HS^-]}{[H_2S]}$$

Eq. 28

Given pK_{eq} of second dissociation = $-\log[1 \times 10^{-7}]$, the pK_{eq} of second dissociation is therefore = 7 and:

$$7 - 14 + pH = \log\frac{[HS^-]}{[H_2S]}$$

Eq. 29

$$-7 + pH = \log\frac{[HS^-]}{[H_2S]}$$

Eq. 30

Therefore, we set up the following table recalling:

Table 1.1. pH influence on the distribution of ions in the second dissociation of Na₂S to base.

pH	$\dfrac{[HS^-]}{[H_2S]}$	$[HS^-]$	$[H_2S]$	% Conversion	Reactants or Product Direction
14	10,000,000	10,000,000	1	99.99999	Reactant
13	1,000,000	1,000,000	1	99.9999	Reactant
12	100,000	100,000	1	99.999	Reactant
11	10,000	10,000	1	99.99	Reactant
10	1,000	1,000	1	99.9	Reactant

9	100	100	1	99	Reactant
8	10	10	1	90	Reactant
7	1	1	1	0	-
6	0.1	1	10	90	Product
5	0.01	1	100	99	Product
4	0.001	1	1,000	99.9	Product
3	0.0001	1	10,000	99.99	Product

Homework

1. If we know [HS⁻], then why are we able to also know [HO⁻]? Show this using the appropriate equation.
2. At which pH in the second dissociation of Na₂S do we have 50% conversion to Reactant? To Product? (Hint: Use Table 1.)

II. Introduction to White Liquor Strengths and ABC Titrations

We are going to learn about how the pulp & paper industry determines what white liquor (cooking liquor) strengths it uses by analyzing the distribution of hydroxide anions from the weak base, sodium sulfide. We will also share the technology to measure these strengths, elucidate the importance of Normality as a key parameter for making proper measurements across the industry, and the universal nature of equivalents.

Let us revisit the first dissociation:

$$\text{Eq. 16}$$

Eq. 22 provided us with $pK_{eq} = (14 - pH) + \log \frac{[Na_2S]}{[HS^-]}$

Knowing $K_{eq} = 10$, $pK_{eq} = -1$ and,

$$-15 + pH = \log \frac{[Na_2S]}{[HS^-]} \qquad \text{Eq. 25}$$

We can then create Table 1.2 (like Table 1.1):

Table 1.2. pH influence on the distribution of ions in the first dissociation of Na₂S to base.

pH	$\dfrac{[Na_2S]}{[HS^-]}$	$[Na_2S]$	$[HS^-]$	% Conversion	Reactants or Product Direction
14	0.1	1	10	90	Product
13	0.01	1	100	99	Product
12	0.001	1	1,000	99.9	Product
11	0.0001	1	10,000	99.99	Product

14

10	0.00001	1	100,000	99.999	Product
9	0.000001	1	1,000,000	99.9999	Product
8	0.0000001	1	10,000,000	99.99999	Product
7	0.00000001	1	1×10^8	99.999999	Product
6	0.000000001	1	1×10^9	99.9999999	Product
5	0.0000000001	1	1×10^{10}	99.99999999	Product
4	0.00000000001	1	1×10^{11}	99.999999999	Product
3	0.000000000001	1	1×10^{12}	99.9999999999	Product

We know that at pH =10, the first dissociation reaction is driven 99.999% to the right which is formation of hydroxide. However, for the second dissociation reaction (see green highlight in Table 1), the reaction is driven 99.9% to the left, which is reactants, not hydroxide! The first dissociation irrespective of pH always goes to product (just a matter of degree), but in order for the second dissociation to go to product (hydroxide production), we need acidic pH (< 7, see Table 1). Thus, to obtain 99.9% production of hydroxide for the second dissociation, what is the pH we need to achieve? pH = 4. At pH = 4, for the first dissociation we get 99.999999999% product conversion. Thus, at pH = 4 we can get both hydroxides at high percentages! We know that the weak base, Na_2S, can product two (2) hydroxides. If we wish to reference the hydroxide producing potential of this weak base in a solution with NaOH (i.e, white liquor), we can say that at pH = 10, we have:

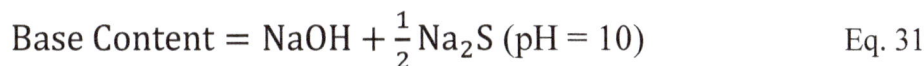

$$\text{Base Content} = NaOH + \frac{1}{2} Na_2S \ (pH = 10) \qquad \text{Eq. 31}$$

Which means all the NaOH dissociates to hydroxide, while only $\frac{1}{2}$ of the Na_2S dissociates to hydroxide (99.999% for first dissociation and 0.1% for second dissociation).

Whereas at pH = 4, we have:

$$\text{Base Content} = NaOH + Na_2S \ (pH = 4) \qquad \text{Eq. 32}$$

Which means all the NaOH dissociates to hydroxide, while the remaining one-half of the Na_2S dissociates to hydroxide (99.999% for first dissociation and 99.9% for second dissociation).

If we wish to use apply these alkalinity formulae (Eqs. 31 & 32) to describe the overall base concentrations, we can do the following:

$$[NaOH] = 2 \times (NaOH + \frac{1}{2} Na_2S) \ [pH \ 10] - (NaOH + Na_2S) \ [pH \ 4] \qquad \text{Eq. 33}$$

$$[Na_2S] = 2 \times (pH \ 4 - pH \ 10) = 2 \times (NaOH + Na_2S - NaOH - \frac{1}{2} Na_2) \rightarrow$$
$$2 \times (0 + \frac{1}{2} Na_2S) \qquad \text{Eq. 34}$$

These formulae (Eqs. 33 & 34) help us to better understand the process of titrations of white liquors. The following information is helpful for what is known as the ABC titration:

Materials Needed:

- **Mill white liquor sample (sodium hydroxide, sodium sulfide, and sodium carbonate)**
- **0.5N hydrochloric acid (HCl) solution**
- **Phenolphthalein indicator**
- **Methyl orange indicator**
- **Distilled water**
- **Burette**
- **Pipette**
- **Erlenmeyer flasks**
- **10% BaCl₂ solution**
- **Formaldehyde**

1. **Sample Preparation:**

   - Pipette a known volume (e.g., 5 mL) of the white liquor sample into an Erlenmeyer flask.

   - Dilute the sample with distilled water to a known volume (e.g., 25 mL).

   - Add 25 mL of a 10% $BaCl_2$ solution to remove any residual sodium carbonate ($BaCl_2$ + Na_2CO_3 → $BaCO_3$ + $2NaCl$) so we have sodium hydroxide and sodium sulfide to measure.

2. **Phenolphthalein Titration:**

   - Add a few drops of phenolphthalein indicator to the diluted sample. The solution should turn pink.

   - Titrate with 0.5N HCl until the pink color disappears. Record the volume of HCl used (V1 or A). The end point will be ~ 9.5 pH (you may use pH meter to measure) which from Eq. 31 will be A (mL) = NaOH + ½ Na_2S.

3. **Convert all of hydrosulfide anion to hydroxide:**

   - Add 10 mL of formaldehyde which will cause the solution to turn pink (which confirms the conversion of hydrosulfide to hydroxide in excess)

   - Continue titrating with 0.5N HCl until the solution turns clear again. Record the volume of HCl used (V2 or B).

4. **Measure the Sodium Carbonate**

   - Add a few drops of methyl orange indicator

   - Add 0.5N HCl until the endpoint (clear solution) which you should record as V3 or C.

Calculations:

1. **Total Alkali (TA) or V3, C (mL) = NaOH + Na₂S + Na₂CO₃**

2. **Effective Alkali (EA) or V1, A (mL) = NaOH + ½ Na₂S**

3. **Active Alkali (AA) or V2, B (mL) = NaOH + Na₂S**

ABC Titration

These measurements give us volume (in mL). We already have the three values for A, B, and C above. If we wish to isolate each of the volumes for each base, how can we do so? Let's start with NaOH. NaOH in terms of A, B, and C can be obtained as follows:

$V_{NaOH} = 2A - B$
$V_{Na2S} = 2(B-A)$
$V_{Na2CO3} = C-B$

Let us run an example to better understand this titration of WL. Say we run the titration as shown on page 13 and get A = 30.9 mL, B = 34.5 mL, and C = 42.0 mL for a 5 mL sample of WL using a 0.5N HCl titrant. In the pulp & paper industry, we want to find the GPL or grams/liter strengths of each of the bases. According to what we have already studied, the endpoint of any titration means:

Equivalents of acid = equivalents of base

L x equivalents of acid/L of acid = L x equivalents of base/L of base

We know from earlier:

N = Number of gram equivalents of solute/Volume of solution in liters.   Eq. 14

Thus,

$$V_{acid} \times N_{acid} = V_{base} \times N_{base} \qquad \text{Eq. 35}$$

Normality relates to the activity of a species using the term "equivalent." A molarity simply expresses the moles of substance available in a 1 L solution, whereas normality takes this concept one step further—it includes its activity in terms of protons or hydroxides (at least in acid/base chemistry) liberated.

If we write the acid equilibrium expression of nitric acid (HNO_3), a strong acid, we recognize it can liberate one proton. Thus, its number of equivalents (proton) = 1.

If we relate this to MW, we get EW = MW/1 (this captures the activity of any acid/base species). Because its MW = 63 g/mole, in a 1 M solution we will have 63 g in 1 L (63 g/L).

Therefore, a 1N HNO_3 solution is equal to EW/L = MW/1/L = 63 g/1/L.

17

Homework

1. You have a 0.5 M solution of hydrochloric acid (HCl). Calculate the equivalent weight of HCl.
2. You have a 1.0 M solution of sodium hydroxide (NaOH). Calculate the equivalent weight of NaOH.
3. You have a 0.25 M solution of sulfuric acid (H_2SO_4). Calculate the equivalent weight of H_2SO_4.
4. What is the Normality of the sulfuric acid solution in the problem above (#3)?
5. You have a 0.1 M solution of phosphoric acid (H_3PO_4). Calculate the equivalent weight of H_3PO_4.
6. What is the Normality of the phosphoric acid solution in the problem above (#5)?
7. Calculate the EW of all the species in a WL sample.

Solution Approach:

1. **Identify the number of replaceable hydrogen ions (for acids) or hydroxide ions (for bases).**
2. **Calculate the molar mass of the acid or base.**
3. **Divide the molar mass by the number of replaceable ions to find the equivalent weight.**
4. **The relationship between molarity and normality is: M x no. of equivalents = N**

III. Titration Measurements

Let us take an example for explaining titration calculations. We want to obtain the GPL for a sodium hydroxide (NaOH) sample which has a volume of 10 mL. We need to calculate the Normality. We will use a 1N HCl titrant, which expends 16 mL to reach the endpoint. Thus, we can use Eq. 31:

V_b = 10 mL (NaOH volume)
N_b = ? (NaOH Normality)
V_a = 16 mL
N_a = 1N
Thus, N_b = (16 mL of HCl) x (1N HCl concentration)/10 mL of NaOH
N_b = 1.6N NaOH concentration

In order to get the GPL, we can use the following relationship:

$$GPL = N \times GPL \text{ (from EW)}/1N \qquad\qquad \text{Eq. 36}$$

Thus, GPL (NaOH) = 1.6N x 40 GPL/1N (knowing EW = MW/1 (1 equivalent of hydroxide)
GPL (NaOH) = 64 GPL of NaOH

Homework

1. You have a WL sample of 5 mL and are using a 1N HCl solution to titrate it to obtain the Na_2S concentration in GPL. Your titration gives the following values: A = 30.9 mL, B = 34.5 mL, and C = 42.0. Provide calculations leading to the final concentration of Na_2S in GPL. Remember the mathematical relationships for all of the WL chemicals (see page 14).
2. Using the same data in 1, please find the final concentration of Na_2CO_3 in GPL.

Na_2O: The "Fictitious Chemical" used to calibrate all WL species

Normalization usually means transforming the data you are working with to a common frame of reference to make "apples to apples" comparisons.

In order to compare/contrast chemical demands and performance on a mill, a chemical standard was chosen: sodium oxide. This chemical is the "normalization" chemical across a mill for every chemical used.

In Homework problem above (#1), we find the solution is 28.1 GPL of Na_2S after all the calculations. If we wish to normalize this to Na_2O, we will need to relate the Normality of this chemical to the chemical we wish to normalize to it. The MW of Na_2O = 62 g/mol. Because there are 2 equivalents of base produced per Na_2O, the EW = MW/2 -= 31 equivalent grams which are equal to 31 GPL per Normality.

Therefore, Na_2S expressed as Na_2O is 28.1 GPL x 31 GPL Na_2O/39 GPL Na_2S (EW of Na_2S) = 22.3 GPL as Na_2O.

To determine the NaOH concentration, we can use V_{NaOH} = 2A – B; we are given A =30.9 mL and B = 34.5 mL so V_{NaOH} = 2(30.9 mL) – 34.5 mL = 27.3 mL of 0.5N HCl (V_a).
To calculate the GPL of NaOH, we first use Eq. 31 (V_{acid} x N_{acid} = V_{base} x N_{base}) to obtain the NaOH concentration:

(27.3 mL of HCl) x (0.5N of HCl)/(5 mL of WL) = 2.73N of NaOH (V_b).

Now, to get the GPL of NaOH, we will take the EW of NaOH = 40.0 g/N (MW/1)/N x 2.73 N = 109.2 GPL of NaOH. In terms of Na_2O, we multiply 109.2 GPL of NaOH x 31.0 GPL of Na_2O/40 GPL NaOH = 84.6 GPL as Na_2O

Titration Curves

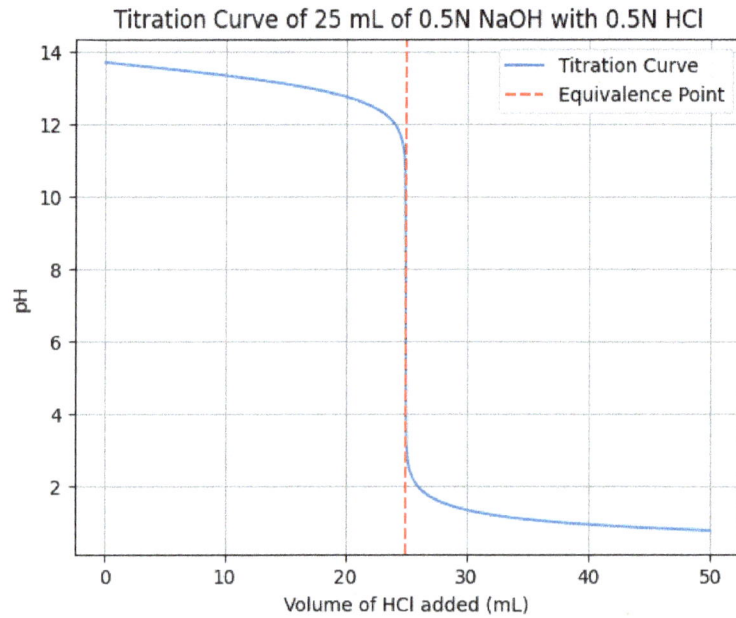

The equivalence (end) point occurs at 25 mL of HCl added, where the pH is neutral (pH 7) because we have used the exact number of equivalents of acid (proton) to neutralize hydroxides in solution. Such an experiment (titration) allows us to calculate the strength of a base solution if it is NOT known.

Now let's run a titration of a weak base, Na_2S:

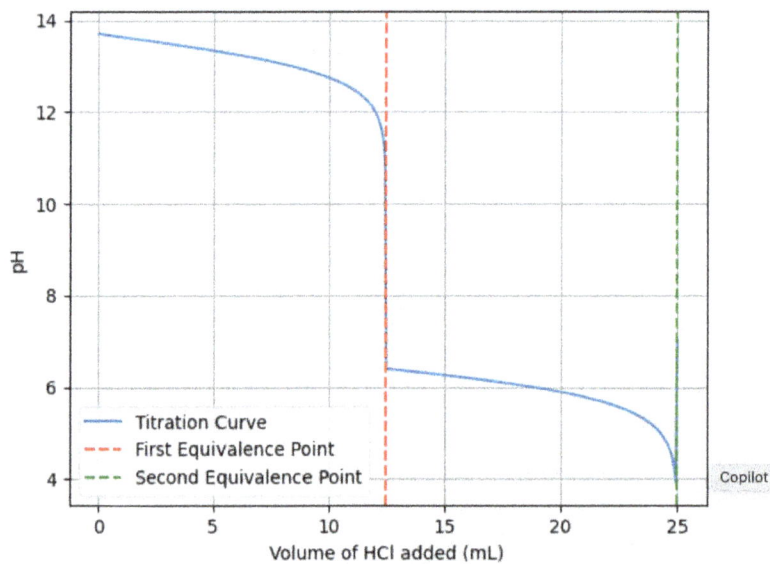

In this titration, there are two equivalence points:

1. The first equivalence point occurs at 12.5 mL of HCl added.
2. The second equivalence point occurs at 25 mL of HCl added.

Homework

1. Sketch the titration curve for the titration of 0.1N HCl with 0.1N NaOH. Indicate the points where the pH is 7 and explain what happens at the equivalence point.
2. Sketch the titration curve for the titration of a solution consisting of 25.0 mL of 0.1N NaOH and 25.0 mL of 0.1N Na_2S using 0.5N HCl.
3. Sketch the titration curve for the titration of 0.2N NaOH with 0.1N NaOH. Indicate the points where the pH is 7 and explain what happens at the equivalence point.
4. Sketch the titration curve for the titration of a solution consisting of 25.0 mL of 0.2N NaOH and 25.0 mL of 0.1N Na_2S using 0.5N HCl.
5. Sketch the titration curve for the titration of a solution consisting of 25.0 mL of 0.2N NaOH and 25.0 mL of 0.1N Na_2S using 0.25N HCl.

Homework Hints:

(1) Use Eq. 35 for all calculations;
(2) You can consider all equivalents in the formula;
(3) You can use Eq. 35 separately for each base and add at the end remembering where sequential titrations occur (pH);
(4) A strong base consumes all acid first regardless of pH.

IV. Redox Chemical Reactions

Reduction and oxidation reactions, like acid and base reactions, are among the most common in all of chemistry. They operate in all aspects of our lives, nature, and in industrial operations. They are particularly evident on the bleaching side of our pulp & paper industrial operations; therefore, for that reason, they merit our undivided attention.

Here is an example of an oxidation:

The oxidation of a cuprous ion (Cu^+) to a cupric ion (Cu^{2+}) can be represented by the following chemical equation:

$Cu+ \rightarrow Cu^{2+} + e-$

In this case, the cuprous ion (Cu+) acts as a reductant (it is a giver of an electron; thinking of red, I think of blood and giving blood).

An oxidant, on the other hand, takes an electron (it is an oxidizer). Common examples in the pulp and paper industry are:

- **Oxygen (O_2)**: Widely used in processes like oxygen delignification (oxygen bleaching) and wastewater treatment.
- **Chlorine dioxide (ClO_2)**: Commonly used for bleaching due to its effectiveness in removing lignin without damaging cellulose fibers.
- **Hydrogen peroxide (H_2O_2)**: Used in bleaching sequences to increase pulp brightness and reduce the use of chlorine-based chemicals.
- **Ozone (O_3)**: Employed in ozone bleaching to achieve high levels of brightness and reduce environmental impact.
- **Peracetic acid (Pa)**: An effective oxidant for bleaching and wastewater treatment.

Again, as in the previous acid/base section, we want to understand the strengths of oxidants such as chlorine dioxide, one of the most common oxidants which is maintained as a liquid. Let us understand how we measure the strengths of oxidants by looking at one of the most common that was used in pulp and paper but serves as a standard: chlorine gas. Here are the steps for the iodometry of chlorine (measurement of its strength):

1. **Sample Preparation**: Pipette a 10 mL sample of aqueous chlorine into an Erlenmeyer flask, add 25 mL of DI water, and add 25 mL of a 10% potassium iodide (KI) to it.
2. **Reaction**: Chlorine in the sample reacts with iodide ions to liberate iodine (I_2) at the same equivalency (1 equivalent of chlorine = 1 equivalent of iodine).
3. **Titration**: Titrate the liberated iodine with a standardized 0.2N sodium thiosulfate ($Na_2S_2O_3$) solution until the yellow color of iodine fades.
4. **Endpoint Detection**: Add a starch indicator near the endpoint. The solution will turn blue, indicating the presence of iodine. Continue titrating until the blue color disappears, indicating the endpoint.

If the total volume of titrant, sodium thiosulfate, is 16 mL (A), calculate the concentration of chlorine in GPL present in the water sample. As stated in the previous acid/base section, we can use Eq. 35, but for oxidant/reductant couples, Eq. 35 becomes:

$$V_{oxidant} \times N_{oxidant} = V_{reductant} \times N_{reductant} \qquad \text{Eq. 37}$$

In this case, $V_{oxidant}$ = 10 mL, $V_{reductant}$ = 16 mL, and $N_{reductant}$ = 0.2N, so $N_{oxidant}$ = 0.32N AND 0.32N of chlorine x EW/1N; what is the EW for chlorine mean? We know the MW = 35.5 g/mol, but how to understand equivalents for chlorine or other oxidants? In other words, what does equivalency mean in redox chemistry? Let us take a deep dive!

For acid/base, equivalency relates to the number of protons or hydroxides released/dissociated.

However, in redox, equivalency relates to the number of electrons released/obtained.

We will need to understand atom valencies to proceed. Why? Because electrons are exchanged which will be indicated by their original → final valency.

Here are the valencies of atoms which will remain constant:

- **Oxygen (O)**: Valency of -2
- **Chlorine (Cl)**: Valency of -1
- **Hydrogen (H)**: Valency of +1
- **Sulfur (S)**: Valency of -2
- **Sodium (Na)**: Valency of +1
- **Calcium (Ca)**: Valency of +2

The redox reaction for the transformation of potassium chlorate ($KClO_3$) to potassium chlorite ($KClO_2$) involves the reduction of chlorate ions (ClO_3^-) to chlorite ions (ClO_2^-). Here are the half-equations for this redox process:

Reduction half-equation: $ClO_3^- \rightarrow ClO_2^-$

We need to know the total transfer of electrons in this case. In chlorate (on left), the valency of chlorine = (3 x -2 per O + n = -1) → -(6) + n = -1 so n = +5 for chlorine in chlorate.

In chlorite (on right), the valency of chlorine = (2 x -2 per O + n = -1) → -(4) + n = -1 so n = +3 for chlorine in chlorite.

Therefore, each Cl transfers one electron/molecule of chlorate. Therefore, the EW = MW/1

Let us go to a reaction using chlorine dioxide. The net reaction of chlorine dioxide with potassium iodide (KI) can be represented as follows:

$ClO_2 + 2I^- \rightarrow Cl^- + I_2 + 2O_2$

In this reaction, chlorine dioxide ions are reduced to chloride ions, and iodide ions are oxidized to iodine. To calculate the equivalent weight of chlorite (ClO_2^-), we need to determine the molar mass and the number of electrons transferred in the redox reaction. The equivalent weight (EW) is given by:

EW=MW/(Number of Electrons Transferred)

Molar Mass of ClO_2^-:

- Chlorine (Cl): 35.5 g/mol (1 atom)
- Oxygen (O): 16.0 g/mol (2 atoms)

Molar Weight of ClO_2 = 35.5 + (2 × 16.0) = 67.5 g/mol

Number of Electrons Transferred: In the reaction, each chlorine in ClO_2 gains 5 electrons to become chloride (Cl^-) because in chlorine dioxide we have (2 x -2 per O + n for Cl = 0 net charge) → n for Cl must be +4 which to get to -1 must gain 5 electrons (it is an oxidant)

Therefore, the equivalent weight of chlorine dioxide is:

EW=67.5 g (MW)/mol/5=13.5 g/equiv

Homework

1. Calculate the EW for MnO_4^- using the following reaction:
   $$2KMnO_4 + 8H_2SO_4 + 10FeSO_4 \longleftrightarrow 2MnSO_4 + 5Fe_2(SO_4)_3 + K_2SO_4 + 8H_2O$$
2. Calculate the EW for $S_2O_3^-$ (thiosulfate) using the following reaction:
   $$I_2 + 2Na_2S_2O_3^- \longleftrightarrow 2NaI + Na_2S_4O_6 \text{ (tetrathionate)}—\textit{this one is very challenging!}$$
3. Calculate the EW for Cl_2 using the following reaction:
   $$Cl_2 + 2KI \longleftrightarrow 2KCl + I_2$$
4. Earlier (page 20) we determined the $N_{ox} = 0.32N$ of chlorine and to get the GPL for its solution, we had to multiply it by EW/1N; From problem #3 above, what is the EW for Cl_2?

V. Titration of Chlorine Dioxide Solution

Chlorine dioxide (ClO_2) is a powerful and selective oxidizing agent widely used in the pulp and paper industry for bleaching. It helps produce brighter and stronger pulp without degrading the cellulose fibers, thus making it a preferred choice over traditional chlorine bleaching methods.

Chlorine dioxide is typically produced on-site at pulp mills due to its instability and the need for immediate use. The most common method involves the reduction of sodium chlorate ($NaClO_3$) with a reducing agent such as methanol, hydrogen peroxide, or sulfur dioxide in an acidic medium. However, chlorate production is contaminated with traces of chlorine.

To determine the strength of chlorine dioxide in grams per liter (g/L), you can use an iodometric titration method. Here's a simplified procedure:

1. **Sample Preparation**: Collect a known volume of the chlorine dioxide solution and start at pH = 7.
2. **Addition of Reagents**: Add an excess of potassium iodide (KI) to the sample. Chlorine and chlorine dioxide reacts with KI to release iodine (I_2).
3. **Titration**: Titrate the released iodine with a standard sodium thiosulfate ($Na_2S_2O_3$) solution until the yellow color of iodine disappears.
4. **Cl_2 Calculation:** Repeat at pH < 7 to allow all ClO_2 and Cl_2 to oxidize excess KI to I_2.
5. Calculate the concentrations of chlorine and chlorine dioxide based on the volumes of sodium thiosulfate used. Chlorine fully reacts at both pHs whereas chlorine dioxide reacts

A (mL) = Cl_2 + 1/5ClO_2 (pH 7)
B (mL) = Cl_2 + ClO_2 (pH 10)

To isolate Cl_2, Cl_2 = ¼(5A – B); to isolate ClO_2, ClO_2 = 5/4(B – A)

Now, to calculate a ClO_2 solution strength, we do as we have done so far. We use the following formula: $V_{oxidant}$ x $N_{oxidant}$ = $V_{reductant}$ x $N_{reductant}$ in which chlorine dioxide is the oxidant and sodium thiosulfate is the reductant. We therefore transform earlier equation 35 to:

$$V_{ClO2} \text{ x } N_{ClO2} = V_{Na2S2O3} \text{ x } N_{Na2S2O3}$$ Eq. 38

Therefore, N_{ClO2} = $V_{Na2S2O3}$ x $N_{Na2S2O3}$/mL of ClO_2 solution

Since ClO_2 = 5/4(B – A) = $V_{Na2S2O3}$ we then have:

N_{ClO2} = 5/4(B – A) x $N_{Na2S2O3}$/mL of ClO_2 solution

And finally, to get the GPL of ClO_2, we take N_{ClO2} and multiply it by ClO_2's EW/N which is 13.5 GPL/N

Homework

1. You have a 100 mL sample of chlorine dioxide solution. After adding an excess of potassium iodide (KI), you titrate the released iodine with 0.1 M sodium thiosulfate ($Na_2S_2O_3$) solution. It takes 1 mL of sodium thiosulfate to titrate the chlorine residual (A, mL) and 25 mL to titrate the chlorine dioxide (B, mL). Calculate the concentrations of chlorine and chlorine dioxide in the original solution in grams per liter (g/L).
2. A 50 mL sample of chlorine dioxide solution is titrated with 0.05 M sodium thiosulfate ($Na_2S_2O_3$). The titration requires 5 mL of sodium thiosulfate for the chlorine residual (A, mL) and 30 mL for the chlorine dioxide (B, mL). Calculate the concentrations of chlorine and chlorine dioxide in the original solution in grams per liter (g/L).
3. A 75 mL sample of a chlorine dioxide solution is then titrated with 0.1 M sodium thiosulfate ($Na_2S_2O_3$). The titration requires 7 mL of sodium thiosulfate for the chlorine residual (A, mL) and 15 mL for the chlorine dioxide (B, mL). Calculate the concentrations of chlorine and chlorine dioxide in the original solution in grams per liter (g/L).

References

[1] 13.1: Introduction to Acid/Base Equilibria - Chemistry LibreTexts (https://chem.libre-texts.org/Bookshelves/General_Chemistry/Chem1_%28Lower%29/13%3A_Acid-Base_Equilibria/13.01%3A_Introduction_to_Acid_Base_Equilibria)
[2] Chapter 4: Acid-Base Equilibrium - Chemistry LibreTexts (https://chem.libre-texts.org/Courses/can/CHEM_220%3A_General_Chemistry_II_-_Chemical_Dynamics/04%3A_Acid-Base_Equilibrium)
[3] Introduction to Acid-Base Equilibria | General College Chemistry II (https://courses.lumen-learning.com/suny-mcc-chemistryformajors-2/chapter/introduction-to-acid-based-equilibria/)

Chapter 2: Wood

The United States is one of the largest producers of pulp and paper globally. As of 2022, it was the largest producer of pulp for paper and the second-largest producer of paper and paperboard, following China. The industry generates significant revenue, contributing billions of dollars to the US economy annually. This includes the production of various paper products, from packaging materials to writing paper. The pulp and paper industry provides employment to many people. In 2023, the industry employed approximately 88,180 workers across various occupations. Jobs in this sector range from production and machine operators to management and administrative roles. Common occupations include paper goods machine setters, operators, and tenders, as well as industrial production managers. The industry offers competitive wages. For example, the mean annual wage for all occupations in the pulp, paper, and paperboard mills sector was around $72,840 in 2023. The industry consists of numerous facilities across the country, including pulp mills, paper mills, and paperboard mills. These facilities are often vertically integrated, meaning they handle multiple stages of production from raw material processing to final product manufacturing. The United States has a high production capacity, with facilities capable of producing large volumes of pulp and paper products. This capacity supports both domestic consumption and export markets. The pulp and paper industry is a vital part of the US economy, providing essential products, significant employment, and substantial economic contributions.

I. Ultrastructure of Wood Fibers

Wood has been a fundamental raw material in the pulp and paper industry for centuries. Its abundance, renewability, and versatility make it an ideal source for producing paper and paper-based products. Chapter 2 will explore the role of wood in the pulp & paper industry, the types of wood used, the pulping process, and the environmental considerations associated with wood sourcing and paper production. We will explore its ultrastructure through the following set of images:

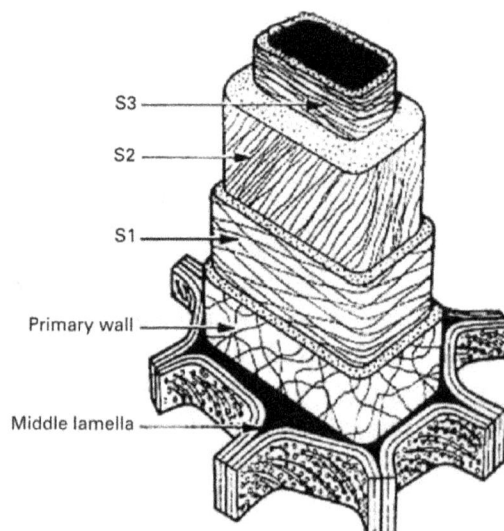

Figure 2.1. A representation of the various sections in the ultrastructure of the wood cell wall.

We will now provide an overview of the various components in the wood cell wall:

Middle Lamella: The middle lamella is rich in pectin and acts as a glue holding adjacent cells together. It facilitates cell adhesion and communication between cells.

Primary Cell Wall (P): The primary cell wall is mainly composed of cellulose, hemicellulose, and pectin to provide flexibility and strength to the growing cell. The cellulose microfibrils are randomly oriented, allowing the cell to expand.

Secondary Cell Wall (S): The secondary cell wall is further divided into three layers: S_1, S_2, and S_3.

- **S_1 Layer:** This layer contains cellulose microfibrils arranged in a helical pattern and provides initial structural support and rigidity to the cell.

- **S_2 Layer:** The S_2 layer is the thickest and contains densely packed cellulose microfibrils aligned in a nearly parallel orientation. It is the main load-bearing layer, providing most of the mechanical strength to the cell wall.

- **S_3 Layer:** The S_3 layer has a similar composition to the S_1 layer but with a different microfibril angle. It provides additional strength and protection to the cell.

Angular Arrangements

S_1 Layer: The microfibrils in the S_1 layer are arranged at a high angle (nearly perpendicular) to the cell axis, typically around 50-70 degrees. This arrangement provides resistance to torsional (twisting) forces and contributes to the overall stability of the cell wall.

S_2 Layer: The middle and thickest layer of the secondary wall. The microfibrils in the S_2 layer are aligned at a much lower angle to the cell axis, usually around 10 to 30 degrees. This layer contains the majority of the cell wall's cellulose and is crucial for the wood's strength and stiffness. The low-angle orientation of the fibers in the S_2 layer provides high tensile strength along the length of the cell, making it the primary load-bearing layer.

S_3 Layer: The innermost layer of the secondary wall, adjacent to the cell lumen. The microfibrils in the S_3 layer are again arranged at a high angle to the cell axis, similar to the S_1 layer. This layer helps in regulating the cell's permeability and contributes to the overall mechanical properties of the cell wall.

Importance in Wood Properties

The varying angles of the microfibrils in the different layers provide a balance of strength, flexibility, and resistance to different types of mechanical stresses. The orientation of fibers affects how chemicals penetrate and react with the wood during pulping. The S_2 layer, being the thickest and most cellulose-rich, is particularly significant in this process.

Molecular Structures in Wood

The **Cellulose Microfibrils** are long chains of glucose molecules linked by β-1,4-glycosidic (glucan) bonds that provide tensile strength and rigidity to the cell wall (see Figure 2.2). As shown in the figure below, among the various molecular species that fill in the microfibrils in addition to cellulose are hemicelluloses, lignin, pectin, and extractives.

• Bundles of β-1,4-glucan chains

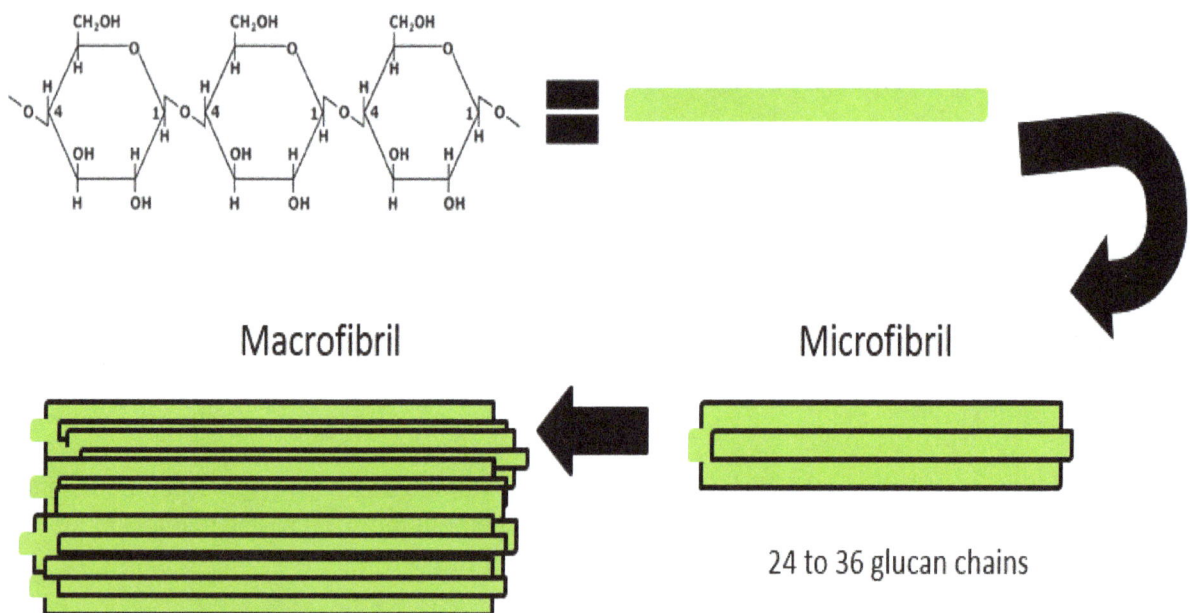

Figure 2.2. A representation of the assembly of bundles of glucan polymeric units stacking to form microfibrils which overlay even further to form macroscopic bundles (macrofibrils).

Hemicelluloses are a heterogeneous group of polysaccharides that bind with cellulose microfibrils (see Figure 2.3 for a representative softwood hemicellulose). It acts as a filler material, providing flexibility and connecting cellulose microfibrils.

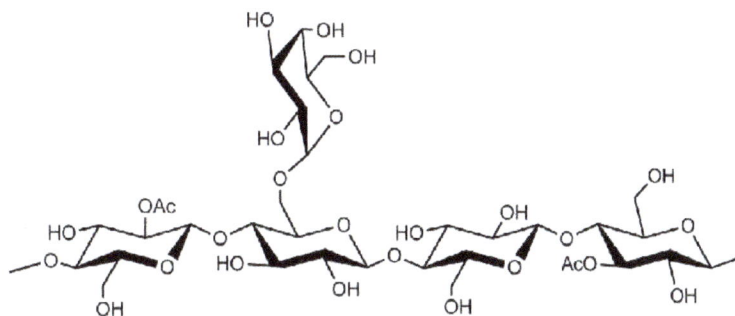

Figure 2.3. A segment of the galactoglucomannan polymer, a common softwood-based hemicellulose.

Lignin is a complex polymer of phenylpropane units that provides compressive strength, rigidity, and resistance to microbial attack (see Figure 2.4).

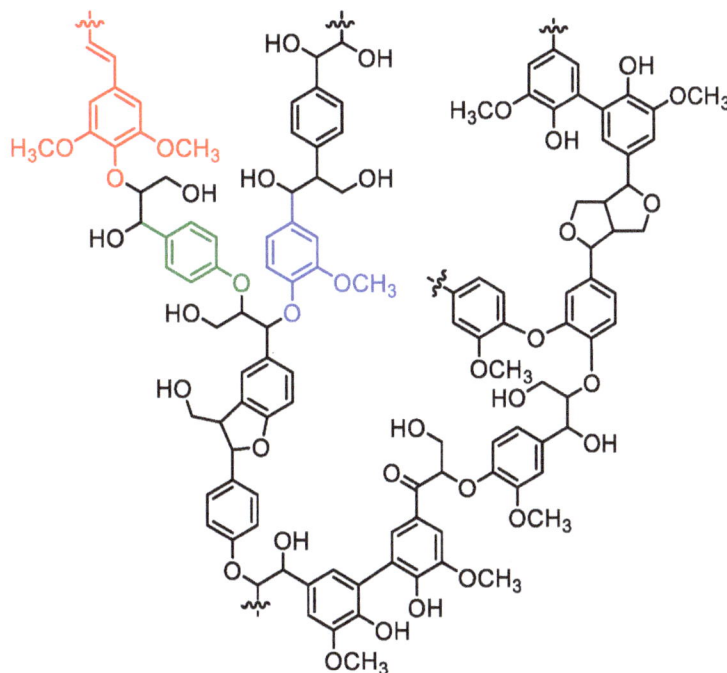

Figure 2.4. A very simplified 2D network depiction of the lignin polymer.

Pectin is a group of polysaccharides rich in galacturonic acid that provides flexibility and porosity to the cell wall, allowing for cell growth and expansion (see Figure 2.5).

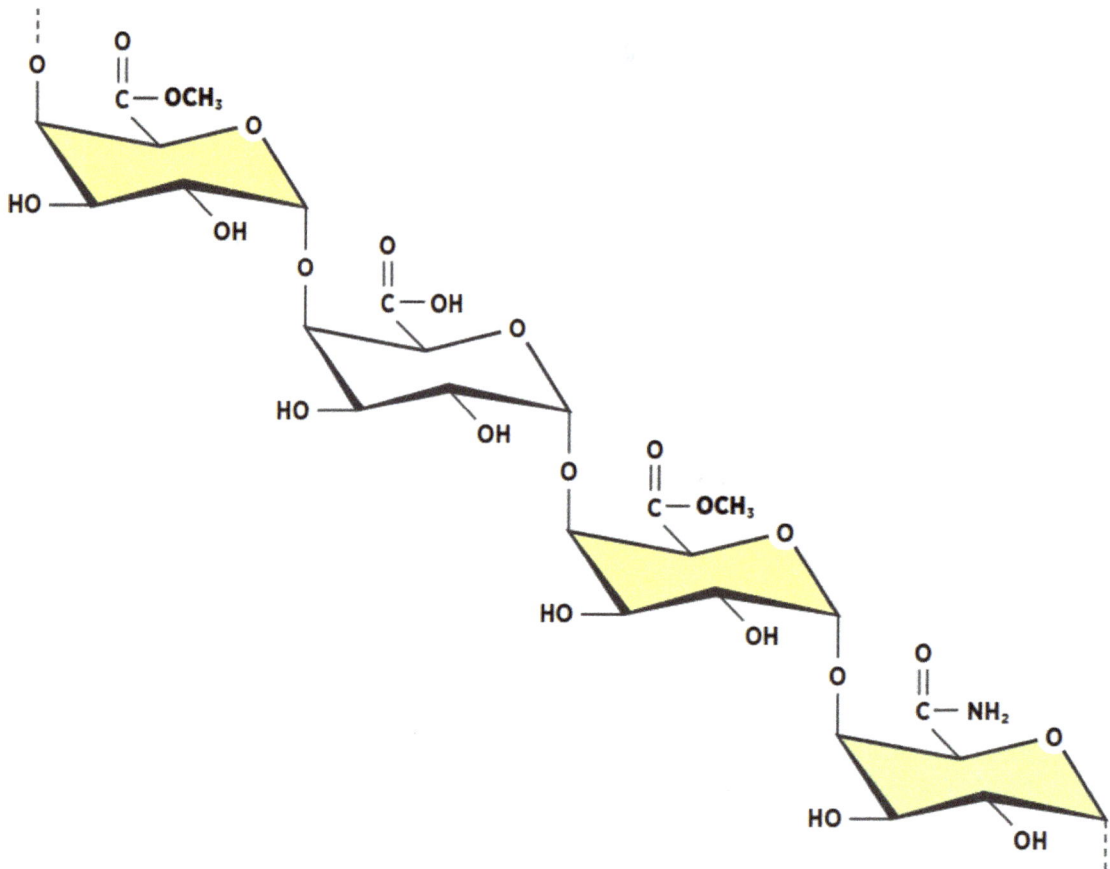

Figure 2.5. A diagrammatic representation of the pectin molecule.

In synergy, these components work together to create a robust and flexible structure that supports the plant's growth and stability.

Extractives in softwoods and hardwoods are organic compounds that are not part of the structural components of the wood, such as cellulose, hemicellulose, and lignin. These extractives can be removed from the wood using solvents and include a variety of substances with different chemical properties. Here are some key points about extractives in softwoods:

- **Resins**: These are sticky substances that can be found in the resin canals of softwoods. They play a role in protecting the tree from pests and diseases.
- **Fatty Acids and Waxes**: These compounds help in reducing water loss and provide a protective barrier against environmental factors.
- **Phenolic Compounds**: These include tannins and lignans, which have antimicrobial properties and contribute to the wood's color and durability.
- **Terpenes**: These are volatile compounds that give softwoods their characteristic scent and can also have antimicrobial properties.

II. Wood Cuts & Macroscopic Components

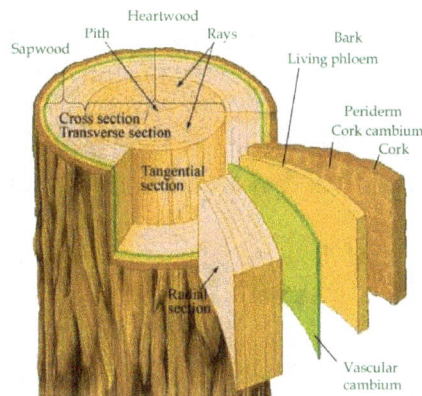

Figure 2.6. A differentiation of the asymmetric wood cuts available in a bolt of wood.

Wood Cuts

Tangential Cut: This cut is made parallel to the growth rings of the wood. It is also known as plain-sawn or flat-sawn. Tangential cuts often display a wavy grain pattern and are prone to warping and cupping as the wood dries. Tangentially cut wood has a higher surface area exposed to chemicals, which can enhance the efficiency of the pulping process. However, the irregular grain can lead to uneven chemical penetration.

Radial Cut: This cut is made perpendicular to the growth rings, running from the center of the log to the outer edge. It is also known as quarter-sawn. Radial cuts produce wood with a straight grain pattern, which is more stable and less prone to warping. Radially cut wood allows for more uniform chemical penetration due to its consistent grain structure. This can result in a more even pulping process and higher-quality pulp.

Axial Cut: This cut is made along the length of the log, parallel to the wood fibers. Axial cuts align with the natural grain of the wood, providing maximum strength and stability. Axially cut wood fibers are more resistant to chemical penetration, which can slow down the pulping process. However, the strength of the fibers can contribute to the production of stronger paper products.

Wood Components

Bark:
- **Outer Bark:** This is the tree's protective layer against external elements like weather, insects, and diseases. It helps retain moisture and insulates the tree.
- **Inner Bark (Phloem):** This layer transports nutrients produced by the leaves to the rest of the tree. It lives for a short period before becoming part of the outer bark.

31

Cambium: This is a thin layer of living cells between the bark and the wood. It is responsible for the tree's growth in diameter by producing new bark and wood cells.

Sapwood: This is the younger, outermost wood that actively transports water and nutrients from the roots to the leaves. It is usually lighter in color and contains both living and dead cells.

Heartwood: Located at the center of the tree, heartwood is composed of older, non-living wood. It provides structural support and is often darker due to the accumulation of extractives.

Pith: This is the small, central core of the tree, consisting of soft, spongy tissue. It is the first part of the tree to form and is surrounded by the initial growth rings.

Rays: These are radial sheets or ribbons extending from the center of the tree to the outer edges. They transport nutrients horizontally across the tree and store food and water.

Phloem: As mentioned earlier, the phloem is part of the inner bark and is responsible for transporting nutrients from the leaves to other parts of the tree.

III. The Role of Wood in Paper Production

Wood is the primary raw material for paper production. As already indicated, it is composed mainly of cellulose fibers, which are essential for creating strong and durable paper. The process of transforming wood into paper involves several steps, including debarking, chipping, pulping, and bleaching. Each step is crucial in ensuring the quality and characteristics of the final paper product.

IV. Types of Wood Used

Different types of wood are used in paper production, each offering unique properties that affect the quality of the paper. The two main categories of wood used are:

Softwood: Softwood comes from gymnosperm trees, which are usually evergreen conifers like pine, spruce, and fir. Softwoods have a simpler cellular structure. They primarily consist of tracheids, which are long, thin cells that transport water and provide structural support. Softwoods lack the vessel elements found in hardwoods, resulting in no visible pores under a microscope. Softwoods grow faster than hardwoods, making them less dense and generally softer. Due to their ease of workability and availability, softwoods are commonly used in construction, paper production, and furniture

Hardwood: Hardwood comes from angiosperm trees, which are typically deciduous trees like oak, maple, and walnut. Hardwoods have a more complex cellular structure. They contain vessel elements, which are large, open cells that transport water. These vessels appear as pores under a microscope and contribute to the prominent grain pattern of hardwoods. Additionally, hardwoods have fibers and parenchyma cells that add to their density and strength. Hardwoods grow slower than softwoods, resulting in a denser and generally harder wood. Hardwoods are often used in high-quality furniture, flooring, and construction projects that require durability and strength.

Key Differences

Density and Strength: Hardwoods are generally denser and stronger due to their complex cell structure and slower growth rate.

Grain Pattern: The presence of vessel elements in hardwoods gives them a more pronounced grain pattern compared to the simpler grain of softwoods.

Workability: Softwoods are easier to work with due to their lower density and simpler structure.

The Pulping Process

The pulping process is the heart of paper production, where wood is broken down into its fibrous components. There are several methods of pulping, including mechanical, chemical, and semi-chemical processes:

1. **Mechanical Pulping**: This method involves physically grinding the wood to separate the fibers. It is energy-intensive but produces high yields of pulp.
2. **Chemical Pulping**: In this process, chemicals are used to dissolve the lignin that binds the cellulose fibers together. The kraft process, which uses sodium hydroxide and sodium sulfide, is the most common chemical pulping method.
3. **Semi-Chemical Pulping**: This method combines mechanical and chemical processes to produce pulp with intermediate properties.

V. Environmental Considerations

The use of wood in paper production has significant environmental implications. Sustainable wood sourcing and responsible forestry practices are essential to minimize the impact on forests and biodiversity. Many paper manufacturers adhere to certifications from organizations like the Forest Stewardship Council (FSC) to ensure sustainable practices. Additionally, recycling paper products reduces the demand for fresh wood pulp and contributes to environmental sustainability.

Homework

1. Wood Density and Strength: Explain how the density of wood affects its strength and durability.
2. Moisture Content: Describe the impact of moisture content on the physical properties of wood. How does moisture content affect the dimensional stability and mechanical properties of wood?
3. Chemical Composition: Identify the primary chemical components of wood and their roles. How do cellulose, hemicellulose, and lignin contribute to the properties of wood?
4. Extractives: Discuss the role of extractives in wood. How do extractives influence the color, scent, and decay resistance of wood? Provide examples of specific extractives and their effects.

5. Wood Cell Structure: Compare and contrast the cell structures of softwoods and hardwoods. How do the differences in cell structure affect the physical properties and uses of these types of wood?
6. Chemical Pulping: Describe the chemical pulping process and its impact on the chemical composition of wood. How do the properties of the wood affect the efficiency and outcome of the pulping process?

Chapter 3: Chemical Pulping

Kraft pulping, also known as the sulfate process, is a widely used method for converting wood into wood pulp, which is the primary raw material for paper production. This process involves treating wood chips with a hot mixture of water, sodium hydroxide (NaOH), and sodium sulfide (Na$_2$S), collectively known as white liquor. The chemicals break down the lignin that binds the cellulose fibers together, allowing the fibers to be separated and processed into pulp. The kraft process is renowned for producing strong and durable pulp, making it ideal for manufacturing high-quality paper products. Additionally, it is more efficient and environmentally friendly compared to other pulping methods, as it allows for the recovery and reuse of chemicals. Kraft pulping is a complex process influenced by several key variables. These variables significantly impact the efficiency, yield, and quality of the pulp produced. Here are the main variables associated with kraft pulping:

I. Terminology

Active Alkali Charge: This refers to the amount of sodium hydroxide (NaOH) and sodium sulfide (Na$_2$S) used in the pulping process. Higher active alkali charges can increase the rate of lignin removal, leading to a lower kappa number (a measure of residual lignin) and higher pulp yield. However, excessive alkali can degrade cellulose and reduce pulp strength.

Sulfidity: Sulfidity is the ratio of sodium sulfide to the total active alkali, expressed as a percentage. Higher sulfidity improves lignin removal and reduces cooking time. It also enhances the selectivity of the pulping process, preserving more cellulose and hemicellulose.

Pulping Temperature: The temperature at which the pulping process is conducted. Higher temperatures accelerate the chemical reactions, reducing cooking time and improving lignin removal. However, too high a temperature can led to cellulose degradation and lower pulp quality.

Pulping Time: The duration for which the wood chips are cooked in the chemical solution. Longer pulping times allow for more complete lignin removal, resulting in a lower kappa number. However, extended cooking can also degrade cellulose and reduce pulp strength.

Wood Species and Chip Quality: The type of wood and the quality of the wood chips used in the process. Different wood species have varying lignin, cellulose, and hemicellulose content, affecting the pulping efficiency and pulp properties. Uniform and high-quality chips ensure consistent cooking and better pulp quality.

Liquor-to-Wood Ratio: The ratio of cooking liquor (chemical solution) to wood chips. A higher liquor-to-wood ratio ensures better penetration of chemicals into the wood chips, improving lignin removal and pulp uniformity. However, it also increases the chemical consumption and cost.

Kappa Number: A measure of the residual lignin content in the pulp. A lower kappa number indicates more complete lignin removal, resulting in higher-quality pulp. However, achieving a very low kappa number can require more chemicals and longer cooking times.

Digester Type and Configuration: The type and design of the digester used for cooking the wood chips. Different digester designs (batch vs. continuous) and configurations can affect the uniformity of cooking, chemical consumption, and overall efficiency of the pulping process.

II. Differences in Pulping Softwoods vs. Hardwoods:

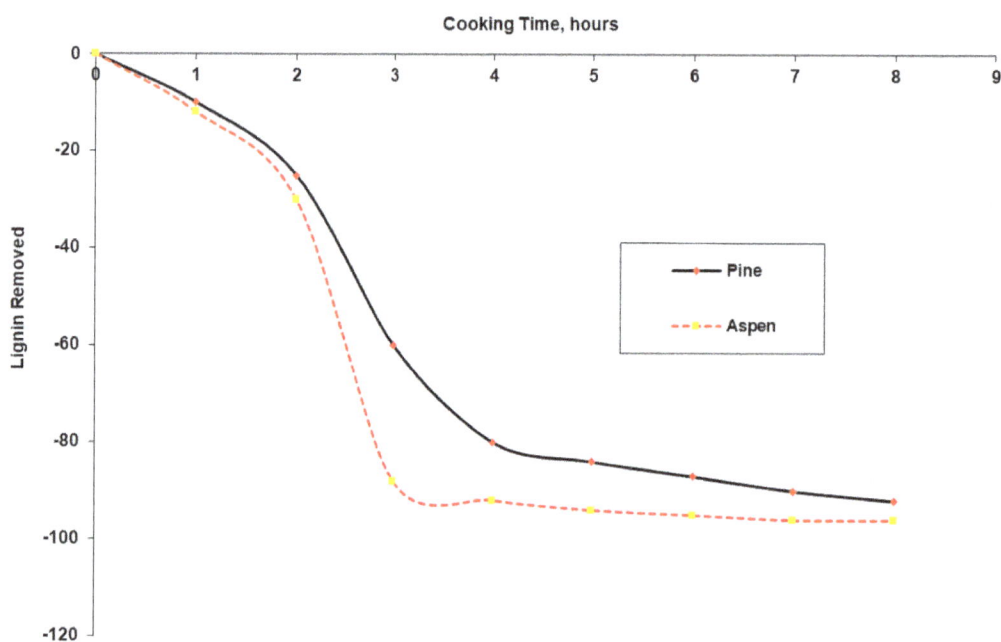

Figure 3.1. A representation of the differences in the pulpability (lignin removed) as a function of cooking time. Courtesy of Dr. Hasan Jameel.

Hardwoods such as aspen (shown in Fig. 3.1) are generally denser than softwoods (pine), which means they have more compact cell structures. They typically have lower lignin content, which requires less time and lower temperatures to break down during the cooking process. Due to their density and lignin content, hardwoods cook more easily than softwoods. This easier cooking rate can result in a more controlled and uniform pulping process, but it also means that hardwoods require shorter cooking times and less energy.

Softwoods are less dense than hardwoods, with more open cell structures. Softwoods generally have higher lignin content compared to hardwoods, making them harder to break down during the cooking process. Softwoods cook slower than hardwoods due to their higher lignin content. This slower cooking rate can lead to protracted pulping processes, but it may also result in less uniformity if not carefully controlled. Cooking hardwoods typically requires less energy due to

36

the shorter cooking times and lower temperatures needed to break down their dense cell structures and lower lignin content. The differences in cooking rates can affect the quality of the pulp produced. Softwoods often yield stronger and more durable pulp, while hardwoods can produce pulp with different characteristics suitable for various applications.

III. Influence of Chip Size on Pulping

Chip size refers to the dimensions of the wood chips used in the pulping process. It includes the length, width, and thickness of the chips. The ideal chip size varies depending on the pulping method and the specific requirements of the pulp mill. Generally, chips should be uniform in size to ensure consistent cooking and chemical penetration.

Cooking Rate: Uniform and appropriately sized chips cook more evenly, leading to a more efficient pulping process. Large chips pulp more slowly than smaller ones, which can result in uneven cooking and lower pulp quality. In Figure 3.2 below, we illustrate the relationship between chip size and cooking efficiency in the pulping process. As chip size increases, the cooking rate decreases, indicating that larger chips are more resistant to uniform chemical penetration and heat transfer. This results in slower and less consistent pulping, which can compromise the overall quality of the pulp. Conversely, smaller and more uniform chips facilitate better chemical access and more even cooking, leading to improved pulping efficiency. The trend underscores the importance of optimizing chip dimensions to enhance process performance and product quality.

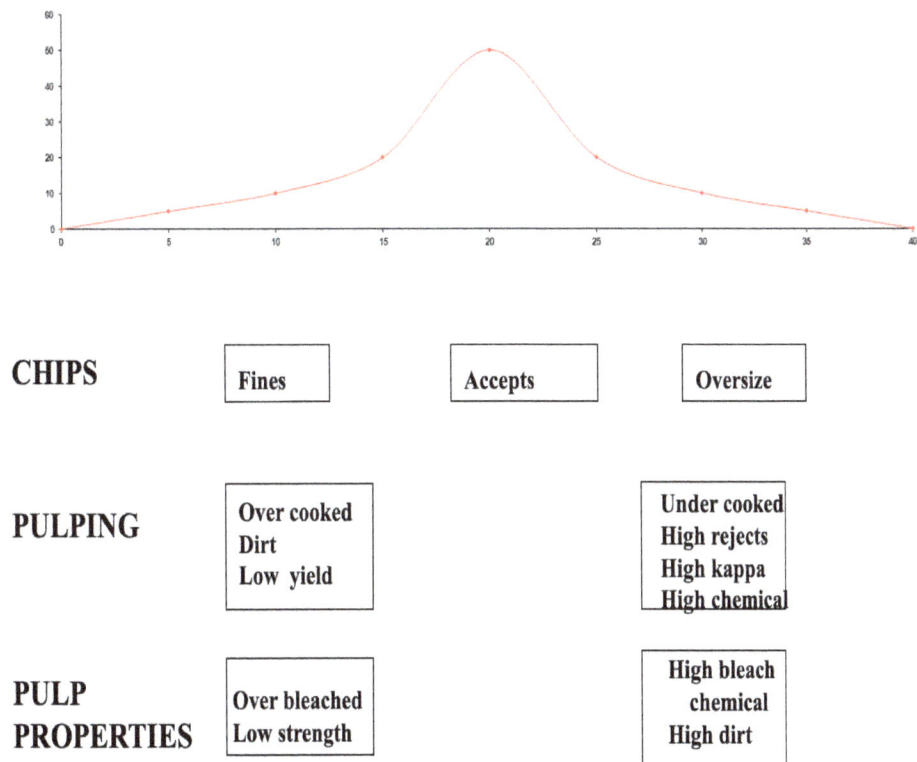

| CHIPS | Fines | Accepts | Oversize |

| PULPING | Over cooked
Dirt
Low yield | | Under cooked
High rejects
High kappa
High chemical |

| PULP PROPERTIES | Over bleached
Low strength | | High bleach chemical
High dirt |

Figure 3.2. A graphical/text representation of how chip dimensions influence pulpability and pulp properties. Courtesy of Dr. Hasan Jameel.

Properly sized chips allow for better penetration of cooking chemicals, ensuring more effective lignin removal and higher pulp yield. Consistent chip size contributes to uniform pulp quality. Variations in chip size can lead to uneven cooking, resulting in a mix of overcooked and under-cooked fibers. Chips that are too small can lead to excessive fines (small particles), which can weaken the pulp. Conversely, overly large chips may not cook thoroughly, leaving behind un-cooked wood that can affect the strength and quality of the final product. Uniform chip size improves the flow of chips through the digester, reducing the risk of blockages and ensuring smooth operation. Properly sized chips can reduce energy consumption during the pulping process, as they require less mechanical and thermal energy to cook thoroughly. Consistent chip size facilitates more efficient screening and cleaning of the pulp after the cooking process. This helps remove unwanted particles and improves the overall quality of the pulp. Uniform chips lead to more consistent refining, enhancing the fiber properties and improving the final paper product. Ensuring the optimal size and uniformity of wood chips is essential for maximizing the efficiency, yield, and quality of the pulping process. Proper chip size management can lead to better chemical penetration, more uniform cooking, and higher-quality pulp, ultimately resulting in superior paper products.

Alkali Charge

Active Alkali (AA): Active alkali refers to the total amount of sodium hydroxide (NaOH) and sodium sulfide (Na_2S) present in the white liquor used for pulping.

$$AA\ (\%) = [([NaOH] + [Na_2S])/\text{Oven-dry wood}] \times 100 \hspace{2cm} \text{Eq. 39}$$

Effective Alkali (EA): Effective alkali is the portion of the active alkali that is actually effective in the pulping process. It primarily includes sodium hydroxide and half of the sodium sulfide.

$$EA\ (\%) = [([NaOH] + 1/2[Na_2S])/\text{Oven-dry wood}] \times 100 \hspace{2cm} \text{Eq. 40}$$

This formula accounts for the fact that only half of the sodium sulfide contributes to the effective alkali charge.

Total Alkali (TA): Total alkali represents the sum of all alkaline substances in the white liquor, including sodium hydroxide, sodium sulfide, and other alkaline compounds like sodium carbonate (Na_2CO_3).

$$TA\ (\%) = [([NaOH] + [Na_2S] + [Na_2CO_3])/\text{Oven-dry wood})] \times 100 \hspace{2cm} \text{Eq. 41}$$

This formula depicts all the alkaline components that contribute to the overall alkalinity of the white liquor.

The balance of these alkali charges affects the delignification process, pulp yield, and quality. Proper control of alkali charges ensures efficient lignin removal while minimizing cellulose degradation. Understanding these charges is essential for the chemical recovery process, as it helps in optimizing the reuse of chemicals and reducing operational costs. By carefully managing the active, effective, and total alkali charges, pulp mills can achieve better control over the pulping process, leading to higher efficiency and improved pulp quality.

Sulfidity

Sulfidity is an important parameter in the kraft pulping process, representing the ratio of sodium sulfide (Na_2S) to the total active alkali (sodium hydroxide ($NaOH$) and sodium sulfide) in the white liquor. It is expressed as a percentage and is crucial for optimizing the delignification process and pulp quality. The formula for sulfidity is:

$$\text{Sulfidity (\%)} = [Na_2S]/([NaOH] + [Na_2S]) \times 100 \qquad \text{Eq. 42}$$

Higher sulfidity improves the selectivity of the delignification process, enhancing lignin removal while preserving cellulose and hemicellulose. Proper control of sulfidity helps in producing stronger pulp by minimizing cellulose degradation. Maintaining the right sulfidity level is essential for efficient chemical recovery and reuse in the kraft pulping process.

IV. Pulping Calculations

Say we are given a white liquor (WL) sample with the following concentrations:

$NaOH$ = 85 GPL as Na_2O, Na_2S = 27 GPL as Na_2O

Active Alkali (AA) is the total $NaOH$ and Na_2O concentrations. Therefore, in this case the AA is equal to 85 GPL ($NaOH$) + 27 GPL (Na_2S) = 112 GPL

Another quantity of great interest to any kraft pulp mill is sulfidity. As we have already been shown, it is equal to Na_2S/($NaOH$ + Na_2S) which in this case is 27 GPL/112GP = 24%

Because %AA is a quantity that specifies the AA within a pulping mixture, we must relate it to the substance (wood chip) we are treating. If, for example, we have a mill specification of 19% AA, then this quantity would necessitate that 19 g of AA relative to 100 g of ODW. Given the WL sample above, how much volume of it would be required to attain a 19% AA?

Volume (WL) = 1L/112 g x 19 g = 0.17 L; this means that 0.17 (170 mL) of WL contain 19 g of AA ($NaOH$ + Na_2S).

Effective Akali (EA) is a more practical description of the "effective" hydroxide which is available at any given time during pulping (remembering that at pulping pH you have all the hydroxides from $NaOH$, but only ½ of the hydroxides from Na_2S. How do we go from AA → EA in this case? Outside of a mathematical relation, which we will extract, let's start from first principles:

We are given that in 0.17 L of this WL sample we have 19 g of AA. However, what is the percentages of each of the substances in the AA? Well, we know that the original WL contained 85 GPL of $NaOH$ as Na_2O and 27 GPL of Na_2S as Na_2O. Thus, in 0.17 L we would have:

85 GPL of NaOH as Na_2O x 0.17 L = 14.45 GPL of NaOH as Na_2O

27 GPL of Na_2S as Na_2O x 0.17 L = 4.59 GPL of Na_2S as Na_2O

Thus, %EA = (NaOH + ½ Na_2S = 14.45 GPL of NaOH as Na_2O + ½ 4.59 GPL of Na_2S as Na_2O)/100 g ODW = 16.75;

Now, to formalize the relationship between EA and AA, we can use the following relationship:

%EA = %AA (1 - %Sulfidity/200)

With respect to modern pulping conditions relative to grades produced, below is a compilation of the alkali charges necessitated for each set of final product and wood type:

	Bleachable	Linerboard
Softwood		
Kappa	30	80-100
%AA	19	16
%EA	16	14
Hardwood		
Kappa	16	
%AA	17	
%EA	14.5	

Notice the kappa (a quantitation of lignin concentration) is addressed by the level of alkali for each grade. In other words, a bleachable grade (one that eventually is fully delignified) will tend to be at EAs ~ 16 for softwood while 14.5 for hardwoods, but "brown" grades (found mostly in softwood lines) tend to be ~ 14.

Homework

1. You have a white liquor sample with the concentrations below. Calculate the active alkali concentration in terms of Na_2O.
   - Sodium hydroxide (NaOH): 100 g/L
   - Sodium sulfide (Na_2S): 50 g/L
2. Given the same white liquor sample as in Problem 1, calculate the effective alkali concentration in terms of Na_2O.
3. Using the white liquor sample from Problem 1, calculate the sulfidity.
4. You mix two white liquor samples below. Calculate the active alkali, effective alkali, and sulfidity of the mixed liquor.
   - Sample A: 150 g/L NaOH and 75 g/L Na_2S
   - Sample B: 50 g/L NaOH and 25 g/L Na_2S
5. In a kraft pulping process, the white liquor has the below concentrations. If the target sulfidity is 25%, determine the required concentration of sodium sulfide to achieve this target.
   - Sodium hydroxide (NaOH): 120 g/L

- Sodium sulfide (Na₂S): 60 g/L

6. A pulp mill reports the following data for their white liquor. Calculate the active alkali, effective alkali, and sulfidity. Discuss how these values impact the pulping process.
   - Sodium hydroxide (NaOH): 110 g/L
   - Sodium sulfide (Na₂S): 55 g/L
   - Sodium carbonate (Na₂CO₃): 30 g/L

V. Yield vs. Kappa

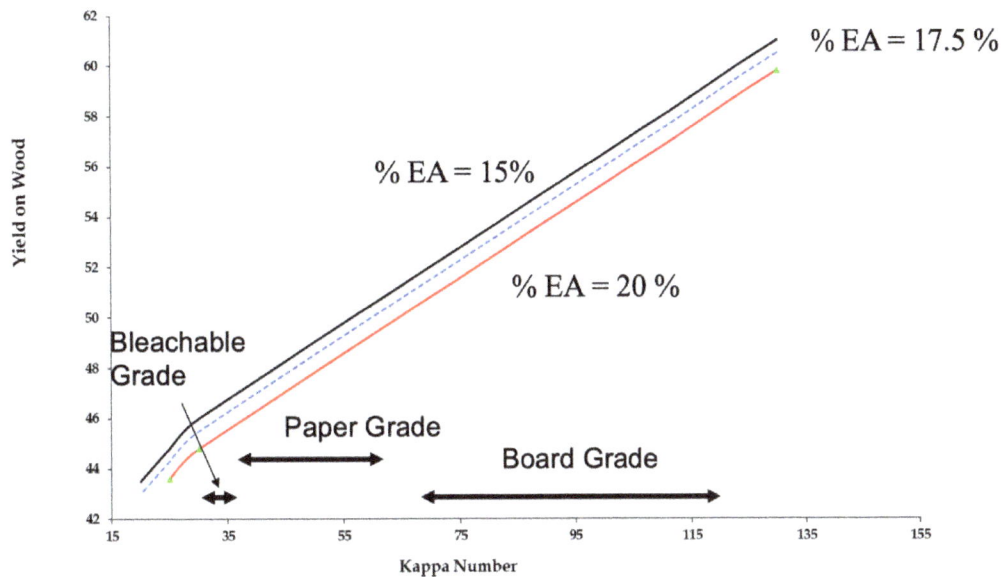

Figure 3.3. A depiction of the relationship of softwood yield on kappa number for a variety of paper grades. Courtesy of Dr. Hasan Jameel.

The above **yield vs. kappa curve** provides valuable insights regarding how efficiently lignin is removed from the wood during the pulping process. A lower kappa number indicates more lignin has been removed, resulting in higher delignification efficiency. As the kappa number decreases as shown in the curve above, the pulp yield typically decreases as well. This is because more wood components, including cellulose and hemicellulose, are dissolved along with lignin which are part of a total yield measurement. The curve helps in balancing the desired pulp quality with an acceptable yield. By analyzing the curve, engineers can optimize the cooking conditions (temperature, time, and chemical concentrations) to achieve the desired balance between yield and kappa number. This helps in producing pulp that meets specific quality requirements while maximizing yield. The curve aids in understanding the economic trade-offs between producing high-quality, low-lignin pulp and maintaining a higher yield. Lower kappa numbers often mean higher chemical and energy costs, so the curve helps in making cost-effective decision. Optimizing the yield vs. kappa relationship can also reduce the environmental impact of the pulping process by minimizing the use of chemicals and energy and reducing waste. In the curves above, notice that lower EAs translate to higher yields while maintaining the same yield vs. kappa slope (no change in the selectivity of pulp losses) which after specific cooking intensities (we will see how this

41

translates to H-factor, next section), we obtain a variety of product grades such as board grade (e.g., liner, sack, kraft paper, etc.) vs. paper grade (i.e., bleachable).

VI. Consumption of Chemicals

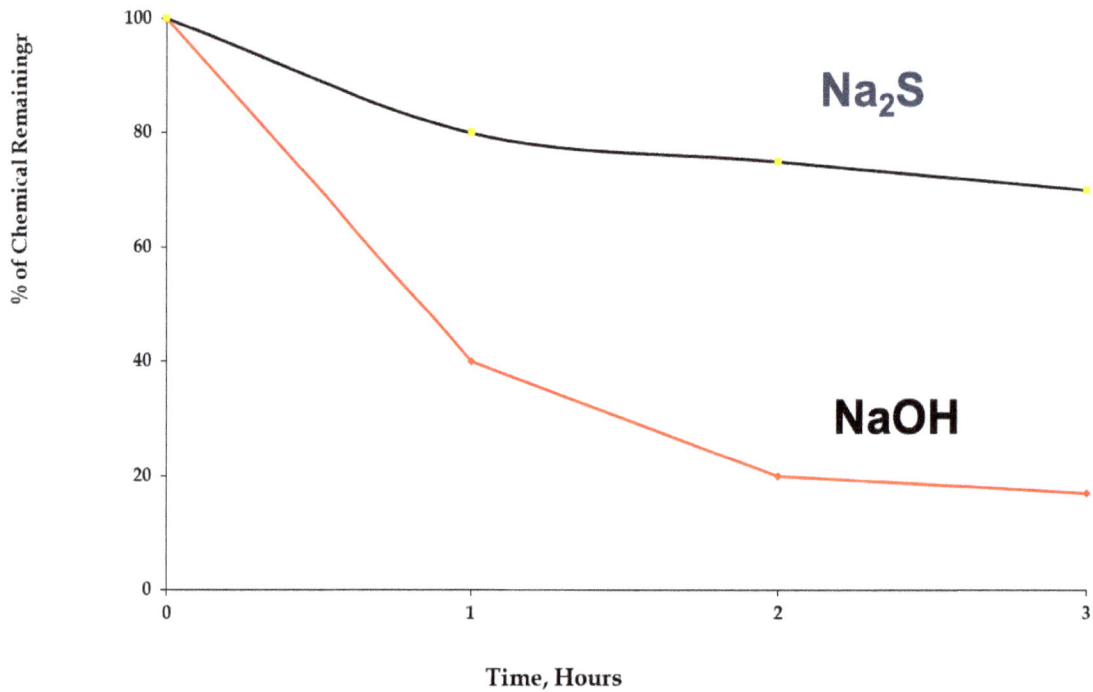

Figure 3.4. An illustration of the rate of white liquor component consumption during the entirety of the process of pulping. Courtesy of Dr. Hasan Jameel.

Shown in Figure 3.4 is the relative chemical consumption of the main cooking chemicals in WL. Notice that the sulfidity is relatively constant throughout the cook whereas sodium hydroxide is expended within three phases. Knowing the above distributions, you could determine the EA (or AA or sulfidity) at any given time if given the starting concentrations.

VII. H-Factor Calculations

The **H-factor** is a crucial concept in kraft pulping, representing a kinetic model for the rate of delignification. It combines the effects of temperature and time into a single variable by assuming that delignification follows a single reaction pathway. Mathematically, the H-factor is calculated by integrating the reaction rate over time, which is influenced by the temperature of the cooking process. This helps in predicting the extent of lignin removal from wood chips during the pulping process. In practical terms, the H-factor allows engineers to optimize the pulping process by adjusting cooking conditions to achieve the desired level of delignification efficiently.

The **H-factor** in kraft pulping is calculated using the following formula:

$$H = \int_0^t e^{\left(\frac{T(t)-T_{ref}}{14.75}\right)} dt$$

Eq. 43

Where:

- H is the H-factor.
- t is the cooking time.
- $T(t)$ is the temperature at time t (in °C).
- (T_{ref}) is the reference temperature, typically 100°C.
- The constant 14.75 is derived from the activation energy for the delignification reaction

The exponential factor is generally captured as a relative rate constant, k, which is depicted through the successive measurements $(T(t) - T_{ref})/14.75$ in a tabular format. This formula integrates the effect of temperature over time, providing a single value that represents the cumulative effect of the cooking conditions on the delignification process ("pulping intensity").

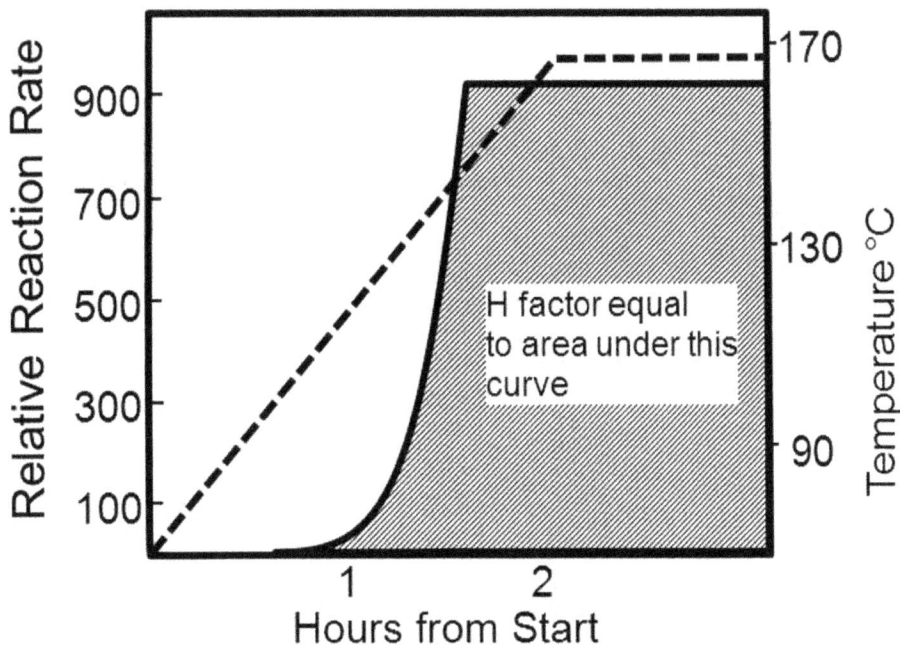

Figure 3.5 A schematic for the relative reaction rate of pulping.

A typical representation of an H-factor graph is shown in Figure 3.5. You can see that there is a region from the start where there is no H-factor of consequence depending on the temperature; for example, at ~1 hour at 110 °C, the H-factor begins to accumulate.

If for example, we are at 170 °C (with relative rate = 921) and 2 hours have elapsed, how much H-factor have we accumulated?

$$\text{H-factor} = kdt = 921 \times 2 \text{ hours} = 1{,}842 \qquad \qquad \text{Eq. 44}$$

Knowing that H-factor is a general intensity factor irrespective of wood source which calibrates delignification (pulping), how long would we have to pulp at 160 °C to reach the same delignification level. This would mean we know the H-factor = 1.842. Using the general H-factor formula (H-factor = kdt), we have:

1,842 (H-factor)/398 (relative rate factor) = 4.63 hours

Therefore, pulping is simply dependent on the lignin content; in fact, a formula was derived expressing this simple dependency on H-factor:

- Kappa No. = $\dfrac{1}{\text{H Factor}}$
- Kappa No. = $196 - 8.49 * \%AA * (1-e^{-H/295})$
  - at constant % Sulfidity and chip quality
- If H Factor and %AA known, can predict Kappa No.

Below in Table 3.1 is the H-factor calculation based on starting at 80°C and ramping over 1½ hours and holding for an additional ¾ hours:

Table 3.1. A listing of the calculations involved in obtaining the total H-factor.

Time (hours)	Temp (C)	k	average k	dt (hours)	H-Factor	Total H-factor
0.00	80	0		0.00		
0.25	95	0	0.0	0.25	0.0	0.0
0.50	110	3	1.5	0.25	0.4	0.4
0.75	125	15	9.0	0.25	2.3	2.7
1.00	140	66	40.5	0.25	10.1	12.4
1.25	155	257.5	161.8	0.25	40.4	50.6
1.50	170	921.4	589.5	0.25	147.4	187.8
1.75	170	921.4	921.4	0.25	230.4	377.7
2.00	170	921.4	921.4	0.25	230.4	460.7
2.25	170	921.4	921.4	0.25	230.4	460.7
				2.3		1,553
				Total Time		Total H-Factor

Thus, we can depict the accumulation of H-factor as a function of time and temperature. The gamut of k's (relative rate constants) is obtained from the chart below:

Table 3.2. A listing of the relative rate constants, k, involved in obtaining the total H-factor according to the calculations listed in Table 3.1.

°C	Relative Rate Constant	°C	Relative Rate Constant	°C	Relative Rate Constant	°C	Relative Rate Constant	°C	Relative Rate Constant
100	1.0	120	9.0	140	65.6	160	397.8	180	2056.7
101	1.1	121	10.0	141	72.1	161	433.4	181	2224.3
102	1.3	122	11.1	142	79.2	162	472.0	182	2404.8
103	1.4	123	12.3	143	86.9	163	513.9	183	2599.0
104	1.6	124	13.6	144	95.4	164	559.2	184	2807.9
105	1.8	125	15.1	145	104.6	165	608.3	185	3032.6
106	2.0	126	16.7	146	114.7	166	661.5	186	3274.2
107	2.2	127	18.5	147	125.7	167	719.1	187	3533.8
108	2.5	128	20.4	148	137.7	168	781.3	188	3812.8
109	2.8	129	22.6	149	150.8	169	848.7	189	4112.5
110	3.1	130	24.9	150	165.0	170	921.4	190	4434.2
111	3.5	131	27.5	151	180.6	171	1000.1	191	4779.6
112	3.8	132	30.4	152	197.4	172	1085.1	192	5150.2
113	4.3	133	33.5	153	215.8	173	1176.9	193	5547.7
114	4.8	134	36.9	154	235.8	174	1275.9	194	5974.1
115	5.3	135	40.7	155	257.5	175	1382.8	195	6431.2
116	5.9	136	44.8	156	281.2	176	1498.1	196	6921.1
117	6.6	137	49.3	157	306.8	177	1622.5	197	7445.9
118	7.3	138	54.3	158	334.7	178	1756.6	198	8008.1
119	8.1	139	59.7	159	365.0	179	1901.1	199	8610.1

VIII. Pulping Calculations

Laboratory-based kraft pulping will be done according to a specific protocol which integrates holistically all the concepts learned to date.

To obtain a bleachable pulp, i.e., ready for chlorine dioxide bleaching, we will employ specific conditions, among which one of the following is:

H-Factor = 1,600
Temperature = 165°C
AA = 18%
Sulfidity = 25%
Chip Weight (**ODW**) = 3,000 g [this is what we want to use]
Liquor/Wood = 4/1
Chip Moisture Content = 45% [chips we obtained from the mill]

We will have two solutions: one of NaOH and the other of Na_2S—this is because sulfidity can be changed as necessary. Upon titration of each solution, we find the following GPLs:

NaOH = 105 GPL as Na_2O
Na_2S = 85 GPL as Na_2O

IX. Steps to Calculating Chip Weight and Volumes of Liquids

1. Let us **FIRST** determine how much water is in the chips. If we divide 3,000 g (ODW) by 55% (solid content, 1- 0.45; 45% = chip moisture content; 55% = 3,000 g/(3,000 g dry chips + water)), then we get 5,454 g of total chips which we must measure. Thus, this amount of chips contains 5,454 g (total weight) – 3,000 (ODW) = 2,454 g of water (chip moisture content; 45% = 2,454 g of water/(3,000 g ODW chips + 2,454 g of water)).

2. **SECOND**, we determine the quantities of NaOH and Na_2S (AA%) we must add: AA% = 18% x OD Weight of Chips (3,000 g) → 18% x 3,000 = 540 g of NaOH + Na_2S

3. How do we calculate the individual chemicals in the WL knowing just the AA? We have the %Sulfidity which is Na_2S/(NaOH + Na_2S). We know it is 25% sulfidity. Thus, 0.25 = Na_2S/(540 g of NaOH + Na_2S) → 0.25 x 540 g = 135 g of Na_2S. Therefore, NaOH = 540 g – 135 g = 404 g.

4. How much Na_2S in volume do we pull from the stock solution which we titrated to be 85 GPL as Na_2O? Because we need 135 g, we divide this by 85 GPL to get the volume in L → 1.59 L of Na_2S stock solution should be pipetted.

5. How much NaOH in volume do we pull from the stock solution which we titrated to be 105 GPL as Na_2O? Because we need 404 g, we divide this by 105 GPL to get the volume in L → 3.86 L of NaOH stock solution should be pipetted.

6. **THIRD**, we need to determine how much liquid to add to 3,000 g OD weight of chips to attain a static 4/1 Liquid (liquor) to OD chip ratio. We can set up the ratio as follows: 4/1 = Liquid Volume/3,000 g OD chips → we need 12,000 g of Liquid (Liquor) to attain this ratio. So, how much volume do we have at this point? We have (1) chip moisture = 2,454 g of water, (2) NaOH volume = 3.86 L, and (3) Na_2S volume = 1.59 L. We know the specific gravity of water (the ratio of the density of a substance to the density of water for a liquid) = 1.0 whereas for the individual chemical solutions = 1.1 g/mL; thus, for NaOH → 3,860 mL x 1.1 g/mL = 4,246 g of NaOH, while for Na_2S → 1,590 mL x 1.1 g/mL = 1,749 g of Na_2S. So for the numerator in the L/W, we have 2,454 g (moisture) + 4,246 g (NaOH) = 1,749 g (Na_2S) = 8,449 g of liquid (liquor); however, we need 12,000 g in the numerator to achieve the correct L/W which means we are missing 12,000 g – 8,449 g = 3,551 g of diluting liquid (either water or black liquor). If we choose water, that means we will need 3,551 mL to fulfill the 4/1 L/W ratio.

7. **FINALLY**, we will add all these quantities, chips first, followed by liquors and water, into a digester which we will pressurize/heat to the proper H-factor.

Homework

1. You have 1000 liters of white liquor. Calculate the amount of oven-dried chips needed to achieve a 4/1 liquor-to-wood ratio.

2. You have 500 kg of oven-dried chips. The effective alkali (EA) is 15% and sulfidity is 25% of a WL solution. Calculate the amount of white liquor needed to achieve a 4/1 liquor-to-wood ratio.

3. You have 1,500 liters of white liquor with an EA of 20% and sulfidity of 30%. You need to dilute it to achieve a 4/1 liquor-to-wood ratio for 400 kg of oven-dried chips. Calculate the final volume of diluted liquor.

4. You have 2000 liters of white liquor with an EA of 25% and sulfidity of 28%. You need to dilute it to achieve a 4/1 liquor-to-wood ratio for 500 kg of oven-dried chips. Calculate the final volume of diluted liquor.

Chapter 4: Bleaching – Terms & Concepts

I. Brief Industrial History

Bleached kraft pulp has evolved over the past 125 years. Today, kraft brownstocks contain 3.0% to 5.5% residual lignin for softwoods *versus* 1.5% to 2.5% for hardwoods. Unbleached brightnesses range between 20% and 30% ISO for softwoods, whereas for hardwoods, it spans between 30% and 45% ISO. Current bleach sequences brighten such pulps from 84% to 92% ISO brightness.

Batch pulp bleaching began at the start of the twentieth century. Calcium hypochlorite powder was added to a Hollander bleacher at low consistency (3% to 4%). Higher pulp consistencies were done with advancing single-stage Beller (*ca.* 7%) and Wolf or VW (15% to 25%) bleachers. Later, molecular chlorine and caustic soda were used to generate sodium hypochlorite. The 1920s to 1930s saw the single-batch hypochlorite stage expand into a continuous three-stage process. Chlorination (C), caustic extraction (E), and sodium hypochlorite (H) stages, coupled with interstage pulp washers, were employed. This increased production rates and lowered chemical costs. Three-stage processes could reach the final brightness target with minimum pulp strength losses.

In the 1940s to 1960s, improvements in generating chlorine dioxide on-site were made. Now kraft pulps could be brighten to 85% ISO or higher using a chlorine dioxide stage (D) in CEHD or CEHDED sequence. Chlorine dioxide was determined to be very selective and adaptable for bleaching when compared with hypochlorite. This eventually led to the CEDED sequence.

From the 1970s to the 1990s, the CEDED sequence evolved. Improvements increased the amount of chlorine dioxide being used in delignification, along with oxygen (O) and/or hydrogen peroxide (P). Chlorine and hypochlorite were eliminated to lower chlorinated organic emissions. Another change that happened was the amount of fresh water that was used around the washers. Open bleach plants in the 1970s used 95 to 155 m^3/t. By using various bleaching filtrate reuse designs and different pulp washing equipment, the amount of fresh water used was reduced to between 25 and 40 m^3/t by the mid-1990s. All of these changes shifted bleach plants from CEDED to DED(EP)D, OD(EO)DD, D(EOP)DD, *etc.*

Ozone usage began in the 1990s in Europe. It slowly expanded in the 2010s to include South America and Asia. Technological advances in pulp mixers made ozonation possible at medium and high pulp consistencies. Low charges (<0.7% ozone on pulp) reduced the residual lignin by Δ0.4% to Δ0.9%. At this delignification level, ozone has a minor impact on paper strength. An ozone stage (Z) occurs after an oxygen delignification stage. It may be followed by caustic extraction or chlorine dioxide. As of the late 2010s, 10 million metric tons of bleached pulp, mostly of eucalyptus varieties, were partially delignified using ozone.

A few other developments occurred during the 2000s to 2010s. One item is the use of bleaching enzymes. Xylanase attacks and hydrolyzes the linkages of the xylans. Hydrolyzing these linkages can assist other oxidants that fragment residual lignin. Such applications can allow for smaller lignin entities to dissolve more easily. A xylan-degrading enzyme is added to the kraft pulp after the decker as it enters brownstock storage. Such treatments can reduce the total amount of chlorine dioxide by 10% to 20%.

Another item is hexenuronic acid (HexA) removal. Such species are formed during kraft pulping from the glucuronic acids. These unsaturated entities, located on the xylans, consume ozone and *in situ* hypochlorous acid formed from chlorine dioxide. HexAs are unreactive with oxygen, peroxide, and chlorine dioxide itself. Such components are removed by a hot acid stage or hot chlorine dioxide stage at 85 to 95°C (185 to 203°F). In a hot chlorine dioxide stage, the oxidant quickly reacts with the residual lignin within the first 20 minutes. Then, the residual acid (2 to 3 pH) removes the HexAs within the next hour or two. A hot acid or hot chlorine dioxide stage can reduce the total amount of chlorine dioxide used for bleaching by 20% to 50%. HexAs found on unbleached or oxygen-delignified hardwoods are much higher than corresponding softwoods. Hence, a hot acid or hot chlorine dioxide stage is more effective on hardwood pulps.

This brief introduction covered some key developments in chemical pulp bleaching. The following chapter will present some pertinent concepts encountered during bleaching. Individual bleaching stages typically encountered in the United States will be discussed in the following chapter.

II. Bleaching in General

So, what is bleaching? Bleaching is defined as removing colored components from the pulp produced *via* the kraft process to increase its brightness, as well as removing other undesired impurities (such as extractives, partially pulped wood and bark). The bleaching process removes such impurities selectively while preserving the pulp's strength so that the integrity of the carbohydrates is preserved and their dissolution is minimized. Such a process is done in a manner that has minimal environmental impact and is sustainable. Bleaching improves the capacity of paper and paperboard for accepting printed or written images.

Bleaching is an important unit operation in pulp manufacturing. Roughly 104 million metric tons of kraft pulp were bleached worldwide in 2022 (Fig. 4.1). Of this amount, about 23.5 million tons were bleached in the United States and Canada. Expansions are occurring in Brazil, Indonesia and China, whereas such productions are waning in the United States and Canada.

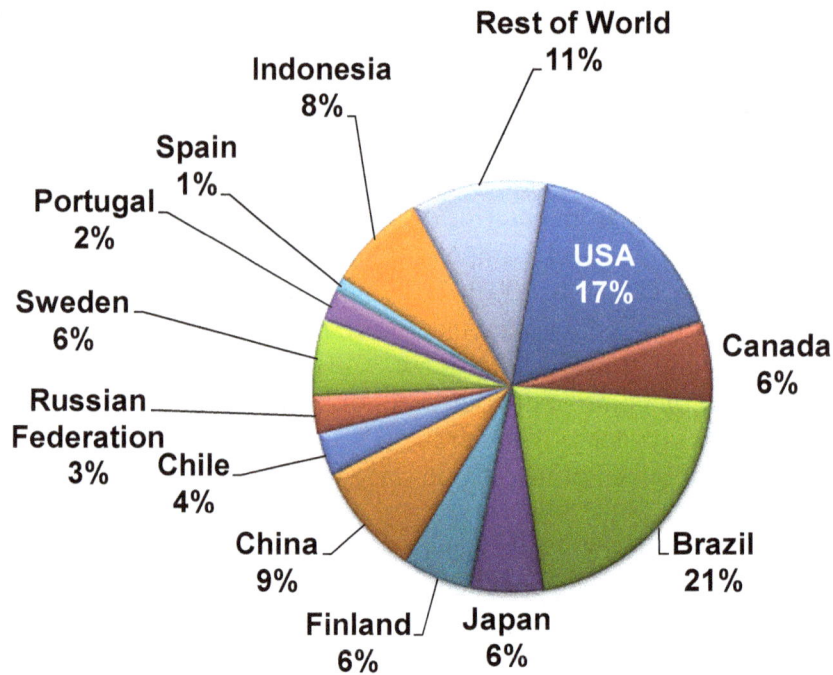

Figure 4.1. Worldwide production of bleached kraft pulp in 2022 (*ca.* 104 million metric tons/yr) among various countries. Data obtained from the RISI database (2025).

Most bleached kraft production in the United States is used on-site, whereas the majority made in Canada is produced as market pulp. In North America, approximately 63% of the bleached kraft is softwood and the other 37% is mixed hardwoods. This trend is reversed in other parts of the world. Among the wood species bleached worldwide in 2018, approximately 41% were various eucalypt species, followed by 35% softwoods and 24% hardwoods.

III. Kraft Brownstock

There are two broad classifications of kraft pulps suitable for bleaching: softwoods (*ca.* 2 to 4 mm long) and hardwoods (*ca.* 0.5 to 1.5 mm long). The typical constituents of such unbleached pulps are shown in Figure 4.2; these components are cellulose, hemicelluloses, and lignin. Softwood brownstocks contain approximately 80% cellulose, 15% hemicelluloses, and 4.4% lignin. Hardwood brownstocks, on the other hand, contain a higher proportion of hemicelluloses and lower levels of lignin. The other category in Figure 4.2 represents residual wood extractives, such as terpenes, tall oil and fatty acids, as well as trace inorganic compounds. The bleaching process serves to further purify the pulp by stripping these non-carbohydrate components.

50

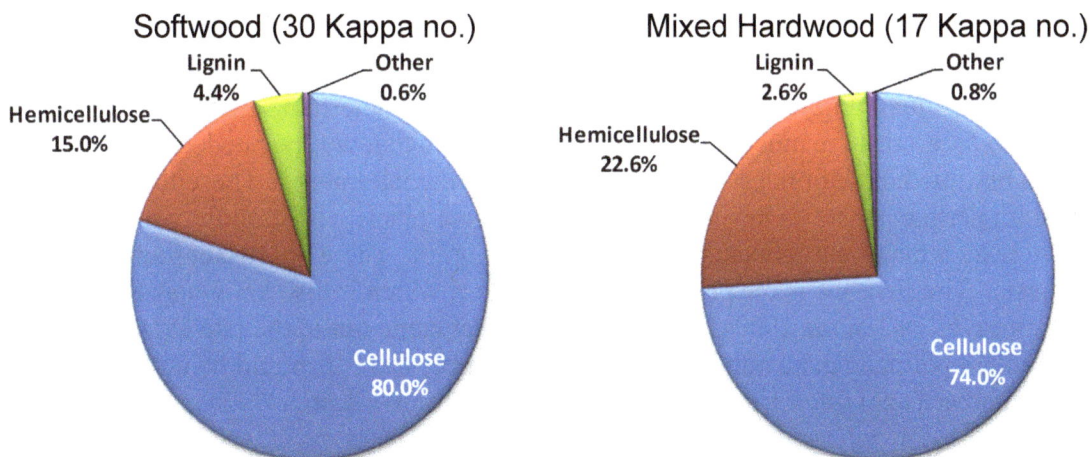

Figure 4.2. Typical kraft brownstock composition for softwood and hardwood pulps prior to bleaching.

Cellulose is the primary component of kraft pulps, and it consists of linear glucose sugar units. It is in higher proportion in unbleached softwoods than in hardwood pulps. It has a high molecular weight and degree of polymerization (700 to 2,000). Cellulose is characterized as being paracrystalline, meaning it has crystalline domains with some amorphous regions. It is colorless to white and is a linear polymer.

Hemicellulose is the second most common polymer component. It consists of various hexose and pentose sugar units. Hardwood pulps have more hemicelluloses than softwoods. Another difference is that hardwood hemicelluloses are mainly glucuronoxylans, whereas softwoods are mainly glactoglucomannans. Hemicelluloses, unlike cellulose, have a much lower molecular weight and degree of polymerization (50 to 300) than cellulose, and the polymer chains are slightly branched. They are white to off-white.

Lignin is the third most abundant polymer found in kraft brownstocks. Native wood lignin is characterized as an amorphous branched heteropolymer made of guaiacyl (one methoxyl), syringyl (two methoxyl), and *p*-hydroxyphenyl units (no methoxyl). Softwood lignins are almost exclusively composed of guaiacyl units, whereas hardwoods are mixtures of syringyl and guaiacyl units, approximately in a 2-to-1 ratio. These phenylpropanes are cross-linked to one another through a variety of saturated carbon-carbon and ether linkages, both from aliphatic and aromatic entities. Weight-average molecular weights of wood lignins are ill-defined, and values noted for kraft pulps are even more so. Reported values are around 20,000 Daltons for softwoods, while lower values are found for hardwoods. Wood lignins are off-white to brown. Kraft residual lignin has been extensively modified during pulping when compared to its native state (<20,000 Daltons). They contain more phenolic end groups (~30% of all units), more saturated and unsaturated carbon-carbon bonds, and fewer ether bonds. This cooked lignin is highly colored, usually dark brown.

IV. Residual Lignin Measurement

The amount of residual lignin found in a bleachable kraft pulp determines how easy it is to bleach. It also ascertains the amount of bleaching chemicals that will be needed. Residual lignin is measured by timed oxidation tests, either in the mill or the laboratory. These time-based tests use potassium permanganate as the oxidant. Permanganate reacts with carbon-carbon double bonds, which are a major component of aromatic lignin units. There are two commonly used chemical tests. The first one is the permanganate number, which is also known as K or P number, and was developed in the 1930s. The second test that came out in the late 1960s as an improvement to the permanganate number is the kappa number test. It should be noted that the permanganate number is not interchangeable with the kappa number test.

The procedure for the permanganate number is given by TAPPI Useful Method UM 251. One gram of oven-dried pulp is mixed with 25 mL of 0.1 N potassium permanganate, 25 mL of 4.0 N sulfuric acid and 700 mL of distilled water. The slurry is allowed to react for 5 minutes at 25.0°C and is ended by adding 5 mL of 1 N potassium iodide. The resulting mixture is titrated with 0.1 N sodium thiosulfate. The permanganate number for the pulp is determined by subtracting the number of milliliters of thiosulfate consumed from 25.

The technique for determining the kappa number is provided by TAPPI Standard Test Method T 236. Approximately 1.5 to 15.0 grams of oven-dried pulp is disintegrated in 800 mL of distilled water. To this slurry are added 100 mL of 0.1 N potassium permanganate and 100 mL of 4.0 N sulfuric acid. This mixture is allowed to react for 10 minutes at 25.0°C. The reaction is terminated by adding 5 mL of 1 N potassium iodide. This solution is titrated against 0.2 N sodium thiosulfate, and the amount of thiosulfate is recorded. An identical test is conducted in the same way but without pulp. The value of p is calculated by doubling the difference between the blank and pulp titrations. If the value of p is between 30 and 70, then the p value is multiplied by a factor given in the T 236 Test Method. This product is then divided by the amount of pulp used to yield the kappa number. If the p-value is less than 30 or greater than 70, then the procedure is repeated by increasing or decreasing the pulp used such that the calculated p is between 30 and 70.

The unbleached kappa number is approximately 6.7 times the percentage of residual lignin in the kraft pulp. (Conversely, the percentage of residual lignin is equal to 0.147 times the kappa number of the pulp). When residual lignin is oxidized during bleaching, such as with chlorine dioxide, this relationship does not remain constant; it changes during bleaching. Although this correlation fluctuates during bleaching delignification, the value still relates how much bleach is required to reach a targeted kappa number or brightness value. The unbleached permanganate number is roughly equal to two-thirds of the kappa number value.

V. Pulp Brightness

An objective of bleaching is to increase the brightness of the pulp. So, what is brightness? To define this parameter, some fundamentals of optics need to be examined. Light interactions with a very thick paper sheet can be explained by the Kubelka-Munk theory. This theory states that the proportion of light that is diffusely reflected from an illuminated surface is governed by two parameters: (1) how much of the light is scattered or reflected, and (2) how much of the light is absorbed. These parameters are denoted as the scattering (s) and the absorption (k) coefficients, respectively. In the Kubelka-Munk theory, the illuminated substrate has sufficient thickness such that the light hitting the sheet is scattered and reflected, and/or is absorbed by the substrate. Light does not pass through the sheet and is lost from the opposite side.

The relationship between the diffusely reflected light at a particular wavelength, R_∞, and the scattering and absorption coefficients is given by this expression:

$$R_\infty = 1 + \frac{k}{s} - \sqrt{\left(\frac{k}{s}\right)^2 + 2\left(\frac{k}{s}\right)}$$

Here, the reflected light R_∞ is expressed as a fractional value between 0 and 1.

When a specific blue light emitted at 457 nm strikes a thick sheet of paper, the measured R_∞ is given a special name brightness. Brightness is expressed as a percentage and is equal to R_∞ multiplied by one hundred. Sheets having a low brightness contain chromophores that absorb this 457 nm wavelength, and therefore the sheet will appear yellow or brown. Those sheets having a high brightness appear white where most all the 457 nm light is reflected to the observer. Brightness may be increased by lowering the light absorption coefficient (color intensity). Or may be increased by increasing the light scattering coefficient. Examples of this include the high brightness of snow when compared to ice cubes, and the high brightness of white sand when contrasted against clear glass.

Another term used with pulp brightness is whiteness. The two terms are *not* synonymous. Pulp whiteness measures light reflections in the red, green, and blue regions of visible light instead of a blue reflection as with brightness. These quantities, known as tristimulus values, are used to calculate a singular whiteness parameter that ranges from 0 to 100.

Little is known about the chromophoric structures in residual kraft lignin. It is conjectured that the residual lignin has similar chromophores as in wood, which is depicted in Figure 4.3 with model lignin compounds. The aromatic rings, when conjugated with carbonyl (C=O) and/or vinyl (C=C) groups, shift the absorptions in the 260 nm UV range to wavelengths into the visible spectra, particularly in the violet, indigo and blue regions. Other structures, such as quinones, have strong absorptions in the 400 to 500 nm and 500 to 600 nm range, which can cause the pulp to appear yellow or red in tint. Such quinones are both destroyed and created during the bleaching process, such as with chlorine dioxide, ozone and oxygen, whereas they are removed by peroxide. Some structures, such as polyphenols, are less colored or nearly colorless but are oxidized to form colored quinones.

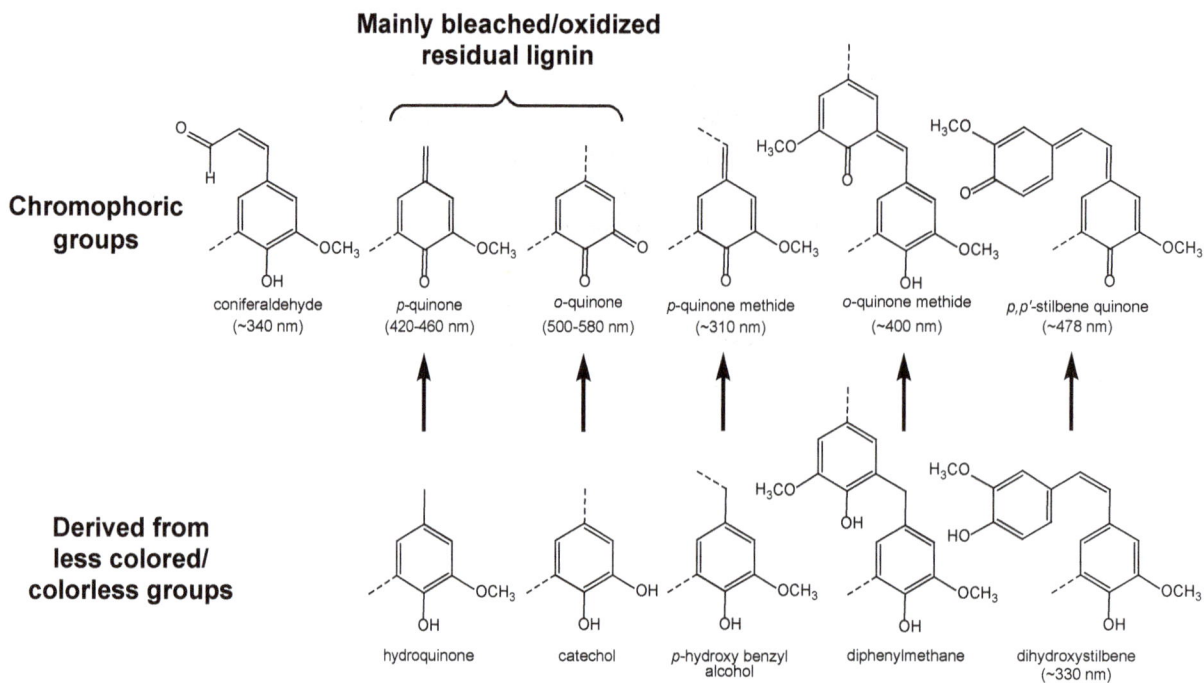

Chromophoric groups

Mainly bleached/oxidized residual lignin

coniferaldehyde (~340 nm)   p-quinone (420-460 nm)   o-quinone (500-580 nm)   p-quinone methide (~310 nm)   o-quinone methide (~400 nm)   p,p'-stilbene quinone (~478 nm)

Derived from less colored/ colorless groups

hydroquinone   catechol   p-hydroxy benzyl alcohol   diphenylmethane   dihydroxystilbene (~330 nm)

Figure 4.3. Suspected residual lignin groups contributing to the light absorption coefficient (*k*) in unbleached or partially bleached kraft pulps.

TAPPI Method T 529 Diffuse Brightness (*d*/0°); "ISO" or "SCAN" Brightness

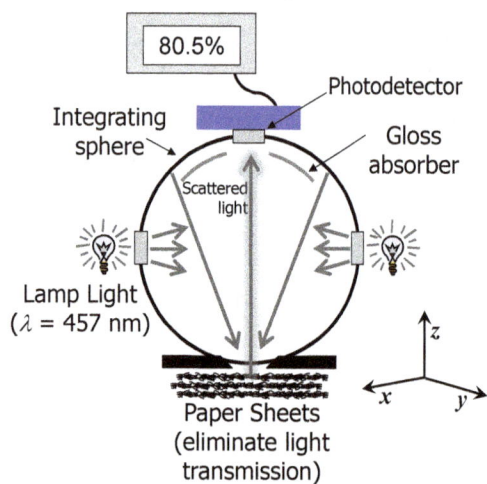

TAPPI Method T 452 Directional Reflectance (45°/0°); "TAPPI" or "GE" Brightness

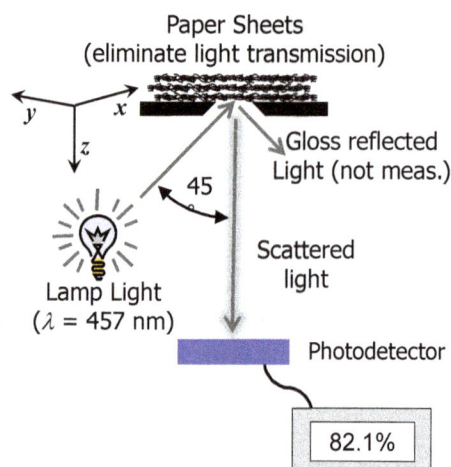

Figure 4.4. Pulp brightness methods: ISO or SCAN (left) and TAPPI or GE (right).

Pulp brightness is measured by two different methods as presented in Figure 4.4. ISO (or SCAN) brightness uses an instrument with the optical geometry shown on the left. TAPPI (or GE) brightness employs a brightness meter illustrated on the right. For TAPPI brightness, 457 nm light strikes the sheet at 45°, and the reflected light perpendicular to the *x-y* plane is measured by the photodetector. Reflected light at angles not 90° to the *x-y* plane is not measured. For ISO brightness, the 457 nm light is bounced off the integrating sphere and is directed towards the opening where the paper sheet is located. This light striking the sheet is diffuse, meaning it

54

strikes the paper at various angles. The light that is reflected normal to the *x-y* plane is measured by the photodetector. In general, ISO brightness is one to two units lower than TAPPI (or GE) brightness for fully bleached handsheets. However, this bias is not valid for measurements made at the papermachine, where fillers, coating and sheet calendaring will affect sheet brightness.

VI. Fiber Strength and Viscosity

Besides residual lignin content and pulp brightness, another parameter of interest during bleaching is fiber strength. Individual fiber strength is related to the length of cellulose polymers that make up the fibers. The cellulose polymers are linear chains, with each unit represented by a link. The degree of polymerization, or DP, represents the average chain length (*i.e.,* the number of linkages) found in the fiber. When the links connecting the glucose units are broken, it lowers the average cellulose chain length. This weakens the strength of the overall fiber if several of them are broken. Cellulose chain cleavage can occur during kraft pulp bleaching. It should be noted that the relationship between cellulose DP and fiber strength properties is not linearly correlated.

Chemical damage to the pulp fibers first manifests itself as a reduction in tear strength. Tear failure of well-bonded softwood sheets involves fiber rupture, and is therefore controlled by fiber strength, not fiber-to-fiber bonding. Tensile (or breaking length) failure comprises fiber pullout and is therefore governed by fiber-to-fiber bond strength. Tear can be thought of how brittle the handsheet is, whereas tensile can be thought of in terms of how much force the handsheet can support before it fails.

It takes several days to evaluate pulp strength properties, from handsheet formation and conditioning to the various strength tests performed. A surrogate method, which only takes a few hours to conduct, is to determine pulp viscosity. Viscosity is measured by dissolving the kraft pulp in a cupric ethylenediamine (CED) solution and observing the time it takes for the solution to pass through a standard capillary. In North America, pulp viscosity is expressed in units of centipoises (cP) or millipascal-seconds (mP·s). This measurement is sometimes referred to as TAPPI Viscosity, which denotes that TAPPI Standard Test Method T 230 was used. In Europe and Scandinavia, pulp viscosity is measured as intrinsic viscosity. Methods such as the Scandinavian Test Method CM15:99 or ISO Standard 5351 describe how this procedure is done. Intrinsic viscosity is expressed as milliliters (or cubic centimeters) per gram, or as cubic meters per kilogram. Either viscosity determination represents an indirect measure of the average length of the pulp's cellulose chains (*i.e.,* its degree of polymerization (DP)).

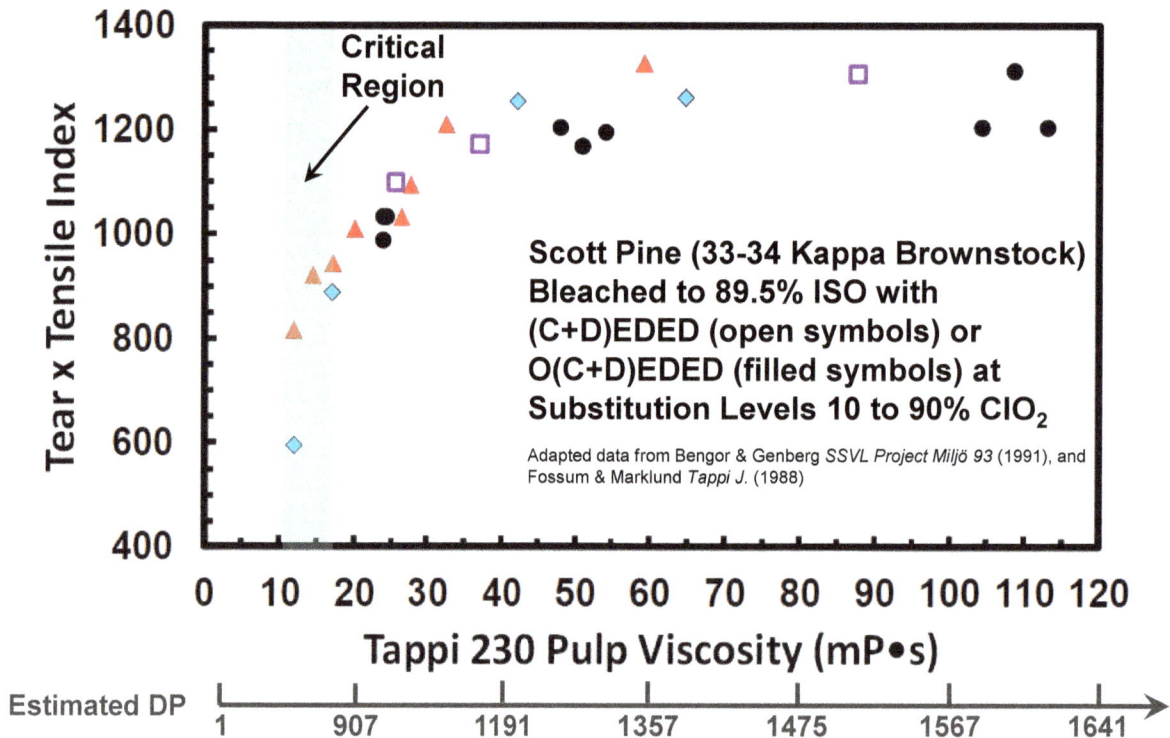

Figure 4.5. Plot of TAPPI T230 pulp viscosity versus tensile-tear index product for bleached softwood kraft pulp.

Pulp strength (at a given tear-tensile product index) is unaffected by changes in kraft softwood viscosity for >30 mP·s when bleached with oxygen, chlorine and/or chlorine dioxide (Fig. 4.5). Strength is proportional to pulp viscosity in the 15 mP·s to 35 mP·s range. It drops precipitously once the viscosity drops below a critical value (~15 mP·s for softwoods). There are some important items to note when making inferences from pulp viscosity. It can be a useful tool to monitor pulp quality during a sequence and can alert when bleaching affects fiber strength as it approaches the critical region. However, viscosity losses must be judiciously analyzed. In this example, a drop in viscosity from 30 mP·s down to 20 mP·s does not correlate to fiber strength loss. Nonetheless, a decrease from 16 mP·s to 10 mP·s correlates to severe strength loss. The second item regards comparing pulp viscosities from different sequences, such as ones that use an ozone stage (Fig. 4.6). Bleaching sequences containing an ozone stage may yield lower viscosity values than a bleach sequence that only uses chlorine dioxide and oxygen. Yet, the same plot as above shows that the highlighted critical region shifts by Δ6 mP·s units. So, caution should be exercised when making inferences about fiber strength from viscosity comparisons between ozone treated pulps *versus* non-ozone treated pulps.

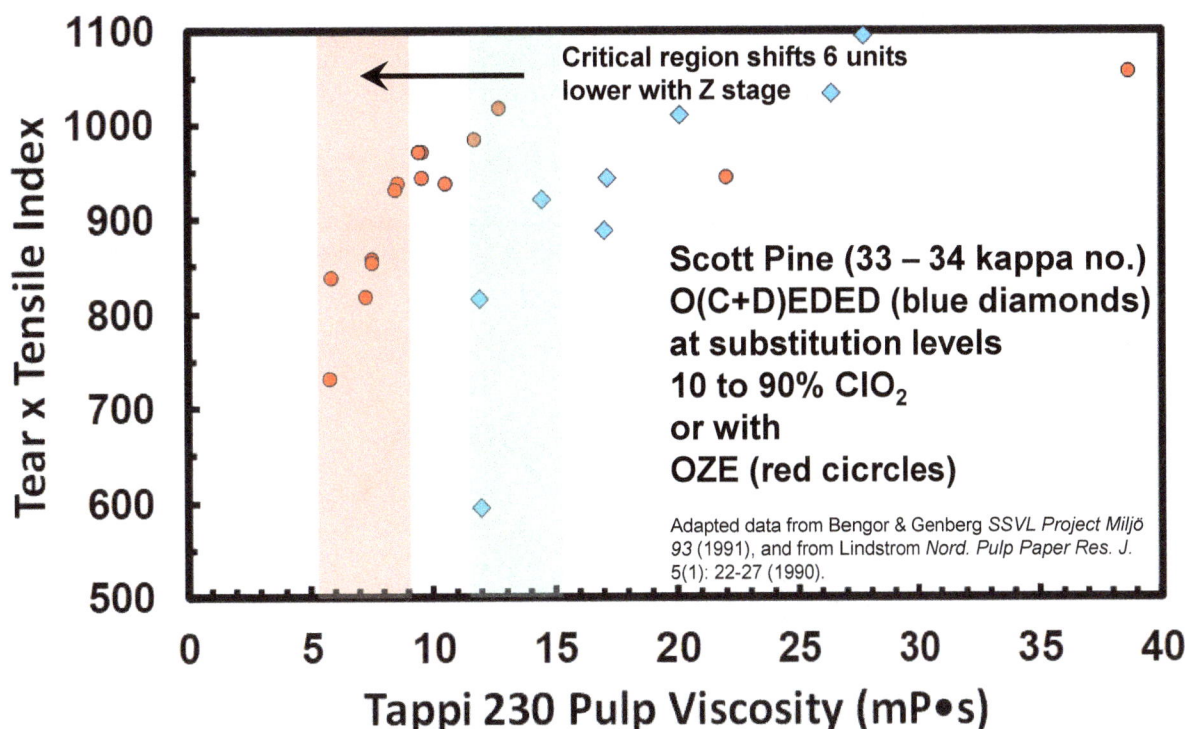

Figure 4.6. Plot of TAPPI T230 pulp viscosity *versus* tensile-tear index product for a softwood kraft pulp delignified with chlorine/chlorine dioxide or with ozone.

VII. Dirt Particles

Dirt particles are unwanted components that are found in kraft pulp. Dirt is defined as dark-colored foreign matter that is not a part of the pulp fibers. It shows up as specks or spots in the paper and has some similarities to residual ink found in recycled pulps. It is quantified as equivalent black area by TAPPI Standard Test Method T 213. Dirt can include impurities that come in with the wood as it is pulped, such as bark, resin, knots (compression wood), and shives (partially pulped wood). Dirt also comes from non-wood sources, such as carbon, sand, equipment rust, rubber, plastic and inorganic scale formation. It can also include grease from equipment and fly ash from pulping.

Ideally, it is best to remove such contaminants from pulp by mechanical methods, such as screeners and centrifugal cleaners, then to rely on bleaching action. Bleaching can remove some dirt contaminates, such as those derived from wood, bark, resins (pitch) and shives. Such chemicals are ineffective in eradicating inorganic dirt, rust and scale.

VIII. Bleaching *versus* Kraft Pulping

Numerous aspects of pulp bleaching are like kraft pulping. Kraft pulping is economical for removing up to 90% of the lignin found in wood to liberate individual fibers. Pulp yield and strength properties remain satisfactory during the kraft process when softwoods are delignified between 20% and 3% residual lignin content on pulp (Fig. 4.7). For hardwoods, pulping can be

57

performed to between 15% and 2.5% residual lignin on pulp. If pulping is forced beyond these lower limits, then pulp yield and strength rapidly decline. Another drawback of the kraft process is that the resulting pulps are dark brown when compared to the original wood, which is light to dark yellow.

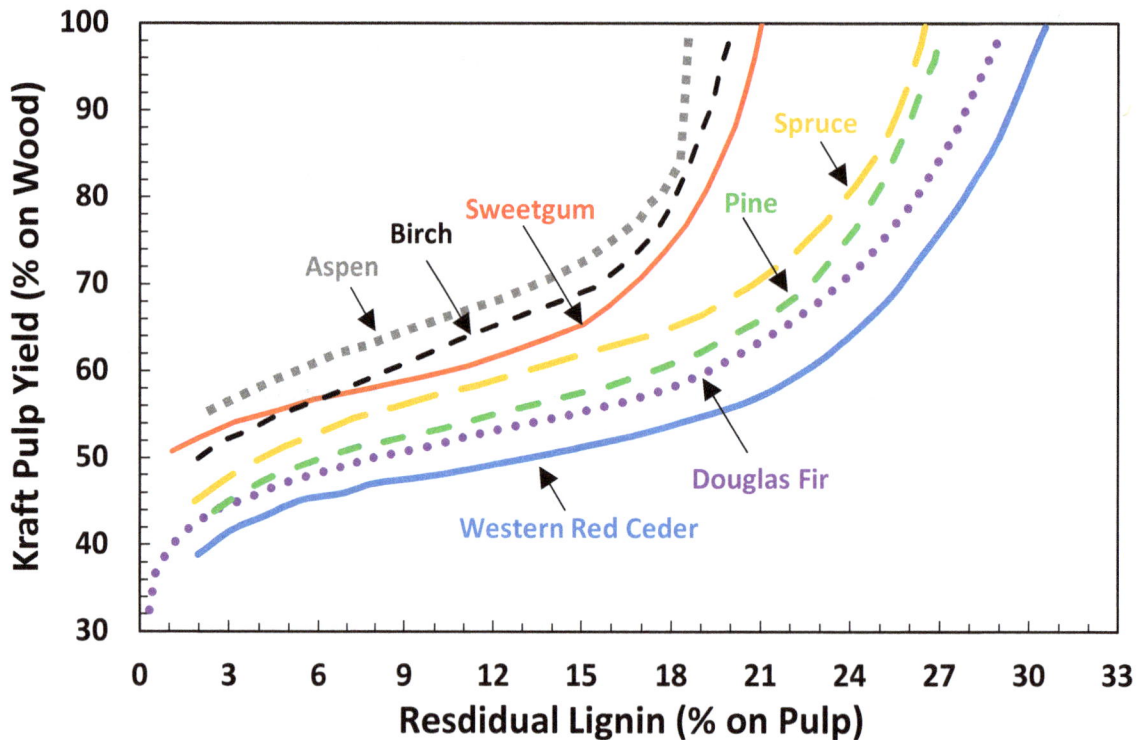

Figure 4.7. Unbleached kraft pulp yields versus lignin content values for various North American hardwood and softwood species.

Chemical bleaching uses oxidants that are more selective at lowering residual lignin or kappa number. Figure 4.8 illustrates some of these aspects for a softwood. The pulp yield drops rapidly when kraft pulping is pushed from 30 to 20 kappa number. This lignin can be removed with less yield loss using oxygen delignification. Oxygen delignification is more selective in decreasing the pulp's kappa number from 25 to 10 in this example. Forcing oxygen delignification to lower than 10 kappa results in steep yield losses. If bleaching technology is used (with or without oxygen delignification), then the pulp's kappa values are reduced to nearly zero with little yield penalty.

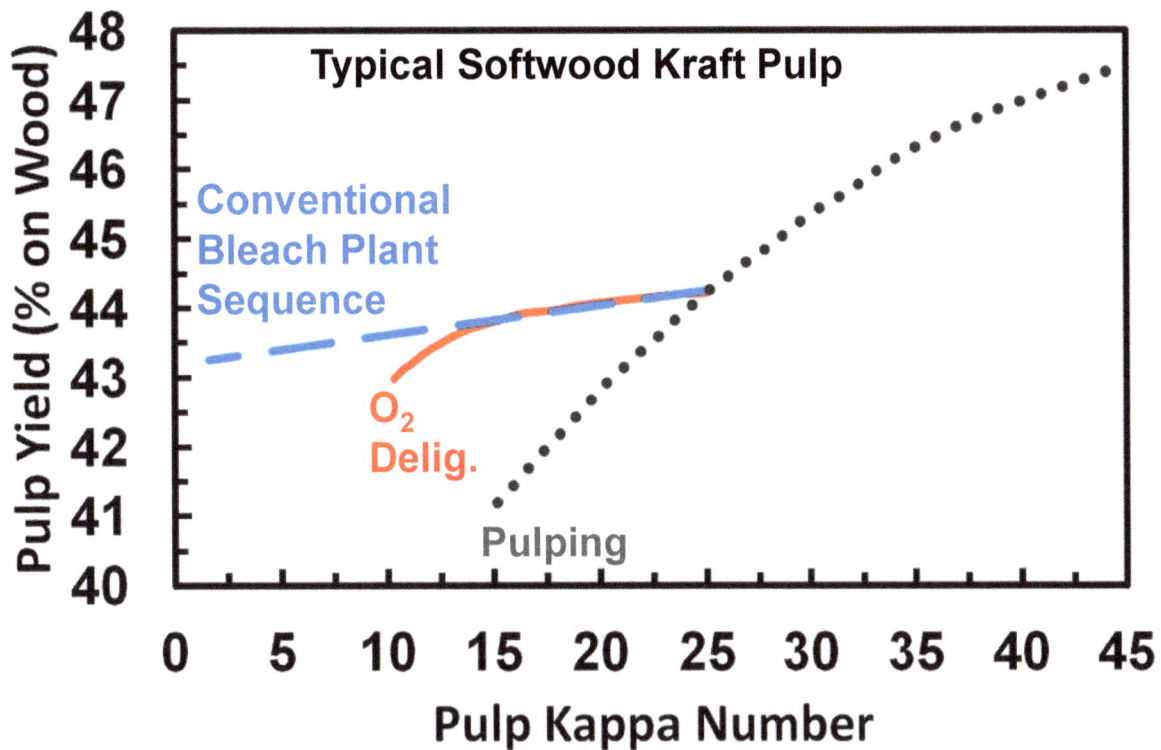

Figure 4.8. Pulp yield *versus* kappa number for a typical kraft pulp, an oxygen-delignified pulp, and fully bleached pulp treated with a D(EOP)DED sequence.

Figure 4.9 shows how bleaching affects pulp viscosity in relation to kappa number reduction for a softwood pulp. As kraft pulping is pressed to lower the lignin content between 30 and 20 kappa, the pulp viscosity drops rapidly. This lignin can be removed with lower viscosity losses using oxygen delignification. Oxygen delignification is selective as the kappa number is reduced from 30 to 16. Delignification levels lower than 16 kappa result in steeper viscosity losses. If bleaching chemicals are used instead, then the kappa number is reduced to nearly zero with minimum viscosity losses.

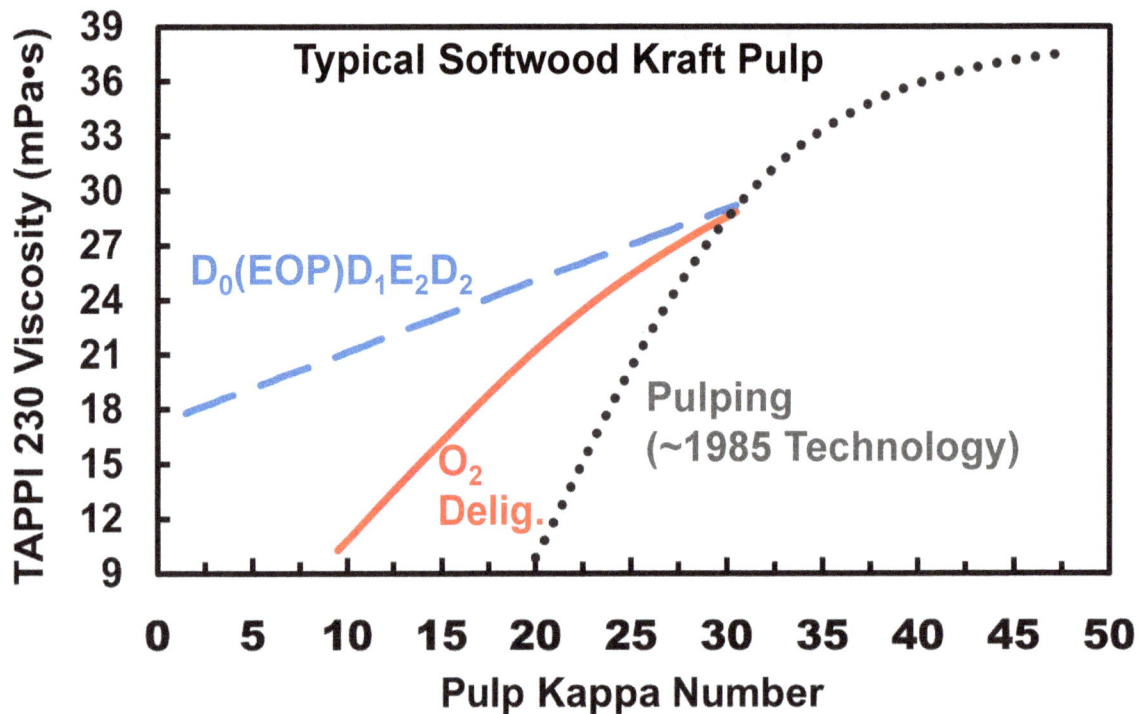

Figure 4.9. TAPPI pulp viscosity *versus* kappa number for a typical kraft pulp, an oxygen-delignified pulp, and bleached pulp treated with a D(EOP)DED sequence.

In the above two examples, it is illustrated that the selectivity of oxygen delignification and/or chlorine dioxide bleaching is higher than extended kraft pulping. Selectivity is defined as the pulp yield or viscosity divided by the kappa number of the pulp. Generally, kraft pulping is ended when the losses in pulp yield or viscosity at a kappa decrease become too great. The drawback of most bleaching oxidants is that they are much more expensive than the sodium sulfide and caustic soda used in kraft pulping. Additionally, the spent chemicals from pulping are easily recovered and reused as opposed to bleaching chemicals, which are used only once. These items are balanced such that kraft pulping is ended when its selectivity is lessened such that oxygen and/or chlorine dioxide bleaching selectivity is higher.

IX. Bleaching Chemicals

Bleaching chemicals break up the hydrophobic and highly colored kraft lignin into smaller and less colored fragments. Some of the bleaching oxidants used today include chlorine dioxide, oxygen, hydrogen peroxide and ozone. Sodium hydroxide (caustic soda) and sulfuric acid are also used to adjust the pulp slurry's pH so that the oxidant can react. In additional cases, caustic is used to solubilize and remove oxidized lignin remains. Other bleaching agents, such as molecular chlorine and sodium hypochlorite, were used in the past but are no longer employed today, except in limited situations in India and China. This is due to these oxidants forming significant chlorinated organic emissions.

Oxidants differ in their ability to remove residual lignin and/or brighten kraft pulp. Terms used to describe how they function include efficiency, reactivity, selectivity and particle removal ability. These descriptors will now be discussed.

Efficiency is defined as the capacity of an oxidant to eliminate lignin or to brighten a pulp without undergoing undesired side reactions or failing to react. Some oxidants are more efficient than others (Table 4.1). Highly efficient oxidants include molecular chlorine, chlorine dioxide and ozone (to a limited extent). Peroxide and sodium hypochlorite are efficient, but not at the level of chlorine dioxide. Peroxide can be catalytically decomposed, and part of its oxidizing potential is lost. Oxygen is less efficient than peroxide. Oxygen reacts with a few lignin functional groups quick, like select phenolics, whereas with other groups, like non-phenolics, it reacts sluggishly or hardly at all.

Table 4.1. Ranking of various oxidants used in kraft pulp bleaching in terms of efficiency, reactivity, selectivity and particle removal.

Bleaching Agent	Efficiency	Reactivity	Selectivity	Particle Removal
Chlorine (Cl_2)	High	High	High	Medium/High
Chlorine Dioxide (ClO_2)	High	Medium/High	High	High
Oxygen (O_2)	Low	Low	Medium	Medium
Peroxide (H_2O_2)	Medium/Low	Low	Medium	Medium
Sodium hypochlorite (NaOCl)	Medium	Medium	Medium/Low	Medium/High
Ozone (O_3)	High	High	Medium	Low

Reactivity is described as the capability of an oxidant to eliminate residual lignin. Ozone is depicted as having high reactivity towards oxidizing and removing lignin, similar to molecular chlorine. Chlorine dioxide has medium-high reactivity, where it takes a longer time to react with lignin (several minutes *versus* seconds). Lastly, oxygen and peroxide are designated as having low reactivity, taking minutes to hours to react.

Selectivity is denoted as the degree to which the oxidant can eliminate lignin or brighten pulps without ravaging the carbohydrate components. In this aspect, chlorine dioxide has high selectivity to remove lignin and to brighten, whereas peroxide has high selectivity to brighten. This is followed by oxygen and then by ozone.

Particle removal ability is represented as the oxidant's capacity to eliminate organic contaminants. By this definition, chlorine dioxide has a high shives removal capacity. This is followed by oxygen and peroxide. Ozone has the least ability to eliminate shives and bark contaminants.

Equivalent weight is used to equate the oxidizing power of various bleaching agents. It is calculated by taking the molecular weight of the oxidant and dividing it by the number of electrons released by the oxidant as it is reduced to its lowest oxidation state. Equivalent or active chlorine is calculated by dividing the oxidant's equivalent weight by the equivalent weight of chlorine (Table 4.2). By this ranking, oxygen and ozone have the highest oxidation potential, followed by chlorine dioxide and peroxide, and then molecular chlorine. However, this oxidant ranking is hypothetical.

Table 4.2. Electron released, equivalent weights and equivalent chlorine for common bleaching agents used in the bleach plant.

Bleaching Agent	Electrons Released	Equivalent Weight	Equivalent Chlorine
Chlorine (Cl_2)	2	35.5	1.00
Chlorine dioxide (ClO_2)	5	13.5	2.63
Oxygen (O_2)	4	8	4.44
Peroxide (H_2O_2)	2	17	2.09
Sodium Hypochlorite (NaOCl)	2	37.2	0.93
Ozone (O_3)	6	8	4.44

X. Bleaching Stages and Equipment

Kraft pulp bleaching is done using a various combination of oxidants and treatments. To help represent such treatment combinations, each bleaching agent is assigned a letter designation. Listed in Table 4.3 is the shorthand nomenclature used to represent various chemical treatments. This is like the periodic table of elements in chemical pulp bleaching. Some rules for this shorthand are given in TAPPI Technical Information Sheet 0606-21.

Table 4.3. Shorthand abbreviations used for various bleaching stages found in kraft pulp bleaching.

Symbol	Stage
C	Molecular Chlorine (Cl_2)
D	Chlorine Dioxide (ClO_2)
Z	Ozone (O_3)
O	Oxygen (O_2)
E	Alkaline or Caustic Extraction (NaOH)
H	Sodium Hypochlorite (NaOCl)
P	Hydrogen Peroxide (H_2O_2)
Pa	Peracetic Acid (CH_3COOOH)
A	Hot Acid Stage
X	Xylanase (enzyme)
Q	Chelation (EDTA or DTPA)

A bleaching stage can use one or more oxidants within a tower or reactor so long as they are compatible with one another. If the oxidants are added sequentially within less than one minute and there is no washing between additions, then the letters are enclosed in parentheses. An example of this is oxygen delignification, where peroxide is added (*i.e.*, (OP)). Another common example is the first caustic extraction stage, E, where both oxygen, O, and peroxide, P, are added (*i.e.*, (EOP)). When compatible oxidants are added with no intra-stage washing, but the timing is greater than one minute, then the letters are not enclosed in parentheses. Instead, the combined letters are separated by a slash. A common example of this is a two-reactor oxygen delignification stage, O/O where the oxygen is added to the pulp several minutes apart. Another example is the high-consistency ozone stage. After two minutes of ozone reaction at high consistency, the pulp is treated afterwards with sodium hydroxide at medium consistency. This is denoted as Z/E.

Often in elemental chlorine-free sequences, or ECF, the bleaching process alternates between acid and alkaline stages. Typically, acidic stages, such as chlorine dioxide (D) and ozone (Z), oxidize lignin and chromophores. A caustic stage (E) solubilizes and extracts the modified lignin and chromophores. Examples of such an alkaline stage include oxygen delignification (O), peroxide stages (P), and first caustic extraction reinforced with either oxygen, peroxide or both ((EO) or (EOP) stage).

A bleach plant uses multiple bleaching stages with different oxidants to remove lignin and decolorize residual chromophores. It is more efficient to use multiple stages than one or two stages. A bleaching sequence can be divided into two parts. The first one to three stages primarily focus on delignification with little to moderate brightening. A common example is the OD(EOP) partial sequence. Chlorine dioxide is often used in multiple places in a bleach plant. To differentiate them in a bleach sequence, numerical indices are sometimes used as subscripts. When chlorine dioxide is used in delignification, it is designated with the 0 subscript.

The last one to three bleaching stages in a sequence focuses on brightening. Here, there is little residual lignin left in the pulp. Chlorine dioxide can brighten the pulp. The first stage that uses chlorine dioxide to brighten is designated by the subscript index 1. The second stage is designated by the subscript index 2. Likewise, if more than one peroxide stage is used, the first one is designated with 1 with the second one designated with 2. Two chlorine dioxide brightening stages may have an intermediate alkaline extraction with (*e.g.*, (EP)) or without peroxide (*e.g.*, E). A special type of chlorine dioxide brightening stage is used in United States bleach plants. Here, caustic soda is added in the dilution zone at the bottom of the D_1 stage, and there is no washing between the chlorine dioxide and the later stage. This is designated as a D_1/E_2 or abbreviated as a D_N stage where the subscript N denotes neutralization at the bottom of D_1.

A bleaching stage is comprised of pumps, mixing devices, a reactor vessel/retention tower, and a pulp washer. Mixers and pumps mix steam and chemicals with the washed pulp, as well as provide momentum to move the pulp slurry. There is a tower or reactor, which provides time for reactions to occur. A reactor may be used if the bleaching process requires pressurization that is higher than the hydraulic pressure (or head), such as with oxygen delignification. Another example is a high-consistency ozone reactor, which requires a specialized pulp mixer. Once the pulp has reacted, it needs to be washed using a washing device. The washer removes dissolved lignin and spent chemicals. Cleaner water is added, and dirty water is removed, which results in bleaching effluent.

Towers typically employ three types of slurry flows. One design is the upflow tower, where pulp is fed at the bottom. This design can be atmospheric or pressurized, as with oxygen delignification. An atmospheric upflow tower is commonly used in the D_0 stage. The hydrostatic head at the bottom of the reactor assists in keeping the chlorine dioxide gas dissolved in the water slurry. Upflow towers use 100% of the tower volume. The retention time is determined by the tower's volume and the production rate. So, if the production rate increases, the retention time is reduced, and *vice versa.*

A second tower design is the downflow tower, where pulp is fed at the top. This design is used with atmospheric stages, as simple extraction, peroxide and peroxide-reinforced extraction stages. This design can allow for flexibility in retention time as the retention time is not coupled to the production rate. The trade-off is that only 70 to 85% of the tower volume is used.

A third tower design is a hybrid that combines the advantages of the previous examples. Up-flow-downflow towers are commonly used with (EO), (EOP), and D brightening stages. The up-flow pre-retention tube has a hydrostatic head, which can assist with keeping oxygen or chlorine dioxide gas dissolved in the aqueous pulp slurry at the beginning of the reaction. The top of the pre-retention tube can have a valve at the top to allow for over pressurization, which can help with (EO) and (EOP) stages.

XI. Bleaching Delignification and Brightening Overview

There are over 90 kraft bleach plant lines in the United States and Canada in the 2020s that bleach to between 85% to 92% ISO. Bleaching to this brightness level is impossible to achieve using a single stage. Multiple stages are needed to brighten the pulp in steps. This is done to maximize the brightness obtained while curtailing the chemicals used. Different oxidant combinations are used to complement one another in delignifying and decolorizing the pulp.

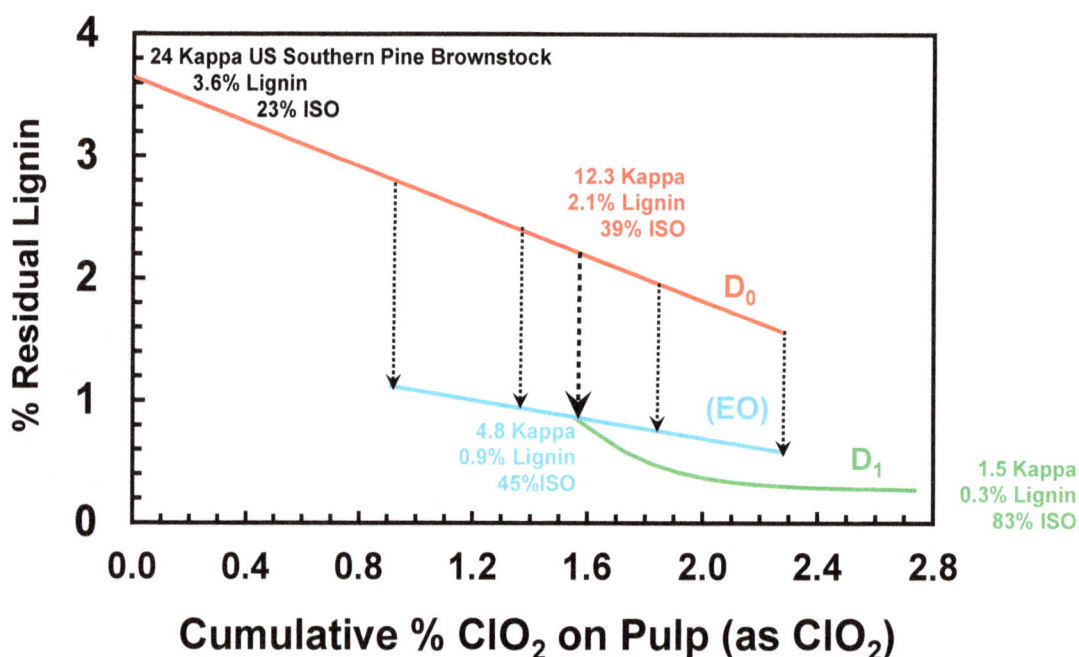

Figure 4.10. Bleaching delignification of a US southern pine brownstock as it progresses through the D(EO)D partial sequence as a function of total chlorine dioxide consumed.

Figure 4.10 illustrates what happens during the first part of the sequence for a US southern pine brownstock. A chlorine dioxide delignification (D_0) stage lowers the residual lignin content from 3.6% down to 1.5%. The brightness increases somewhat from 23% to 39% ISO at 2.1% lignin content. An oxygen-reinforced extraction (EO) stage reduces the lignin content to 0.9%

(4.8 kappa number). A chlorine dioxide brightening (D_1) stage results in a small reduction of lignin content (0.3%) but increases the brightness from 45 to 83% ISO.

Figure 4.11 shows the brightening response of the same brownstock during the $D_0(EO)D_1E_2D_2$ sequence. There is a small rise in brightness from 23% ISO to levels between 40% and 47% ISO after $D_0(EO)$. Much larger brightness gains of 35 or 40 points are realized after the D_1 stage. A few extra brightness points are attained if the $D_0(EO)D_1$ pulp is treated by the E_2D_2 partial sequence (87% to 90% ISO).

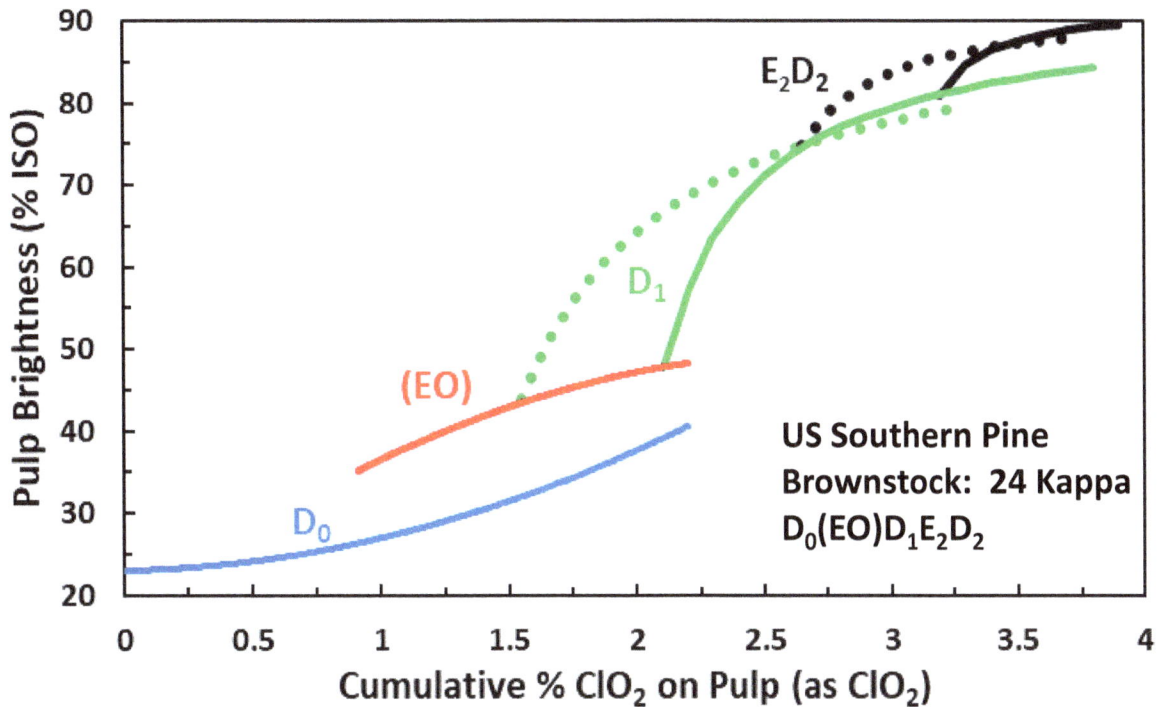

Figure 4.11. Brightening response of a US southern pine brownstock as it progresses through the D(EO)DED sequence as a function of total chlorine dioxide consumed.

Changes in the absorption coefficients of a radiata pine kraft pulp as it progresses through the $OD_0(EO)D_1$ sequence are shown in Figure 4.12. Absorption coefficient values (k) shift as the 30.2 kappa brownstock (4.4% residual lignin) pulp is treated with oxygen delignification and various bleaching stages. Residual lignin has strong absorption coefficients in the ultraviolet range (200 to 320 nm) with some absorbances in the visible spectrum (400 to 700 nm). The bleaching process lowers the k values. Kraft pulp bleaching does not affect the light scattering coefficient (s value). The vertical line marks where brightness measurement is conducted. The k value at 457 nm is decreased from 24.5 m^2/kg for unbleached, to 20.0 m^2/kg after oxygen delignification, to 9.3 m^2/kg after D_0, to 4.6 m^2/kg after (EO), and to 0.6 m^2/kg after D_1. The k values over the ultraviolet region are also reduced during $OD_0(EO)D_1$ sequence but remain high (10 to 60 m^2/kg).

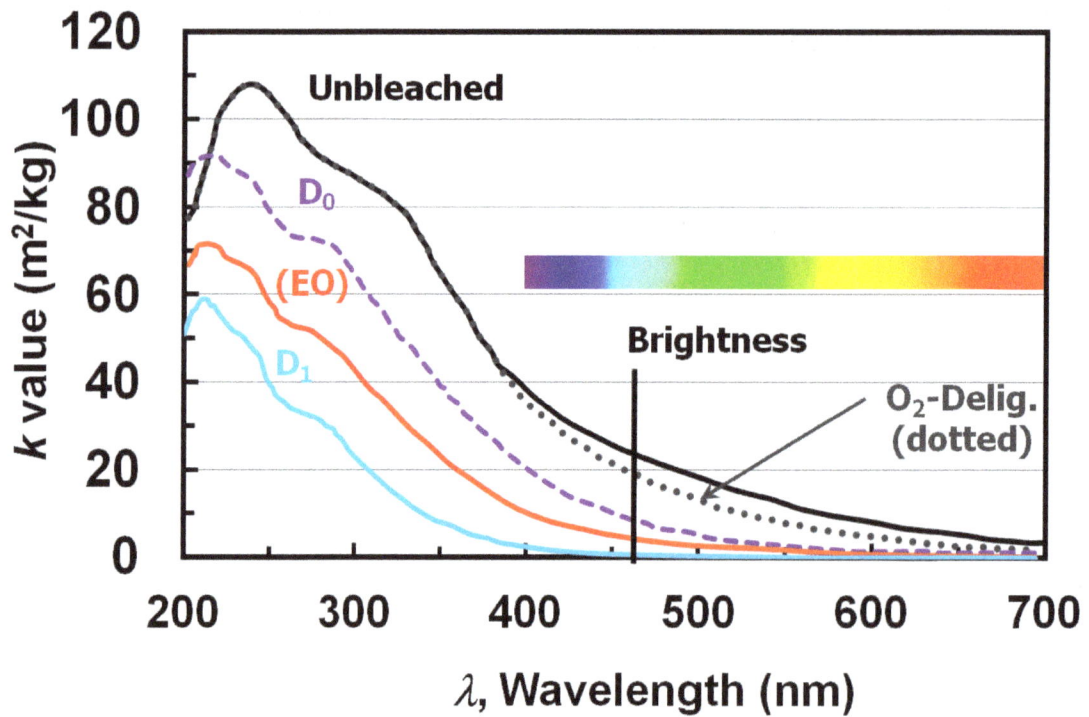

Figure 4.12. Effects of the OD(EO)D sequence on the light absorption coefficient (k) for a radiata pine kraft pulp at various wavelengths.

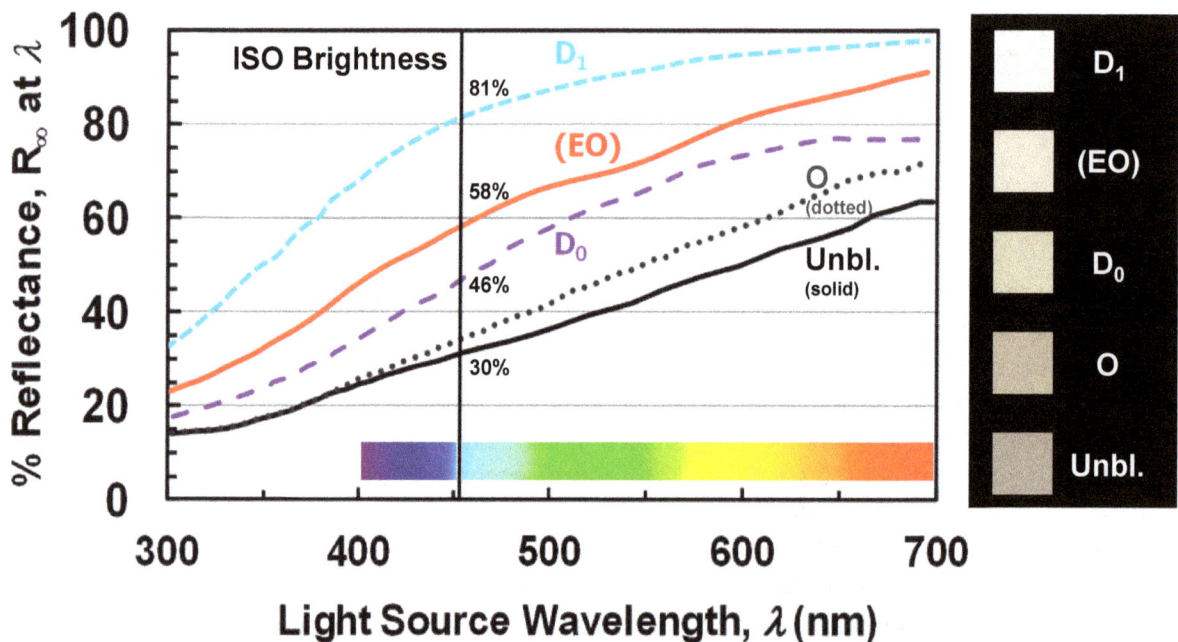

Figure 4.13. Effects of the $OD_0(EO)D_1$ sequence on the reflectance (R_∞) percentage of a radiata pine kraft pulp at various wavelengths.

The corresponding R_∞ values from 300 nm to 700 nm are provided in Figure 4.13 for the radiata pine pulp. The upright line notes where brightness is measured (*i.e.*, 457 nm). As the pulp goes through the $OD_0(EO)D_1$ sequence, more and more light is reflected. When more light is reflected, the surface will appear to be whiter. The right side of the graph represents the pulp's visual appearance. The brightness increases from 30% ISO for the unbleached pulp to approximately 81% ISO after the D_1.

The overall bleaching response of kraft hardwood pulps has notable differences when compared to its softwood counterparts. Figure 4.14 depicts the brightening of an oxygen-delignified eucalypt pulp (59.6% ISO) by the $D_0(EO)D_1E_2D_2$ sequence. Low levels of chlorine dioxide added to the D_0 stage can increase the brightness from 59.6% ISO to between 72% and 82% ISO while reducing the lignin level from 1.0% to 0.8% (4 to 5 kappa no.). This level of delignification makes the D_0 stage of eucalypt kraft pulp behave more like a D_1 stage for a pine kraft pulp. The D_1 stage increases the brightness of the $OD_0(EO)$ pulp by 10 to 12 points to levels of 87% to 88% ISO. The E_2D_2 partial sequence can brighten the eucalypt pulp by a few points to reach 89% to 91% ISO.

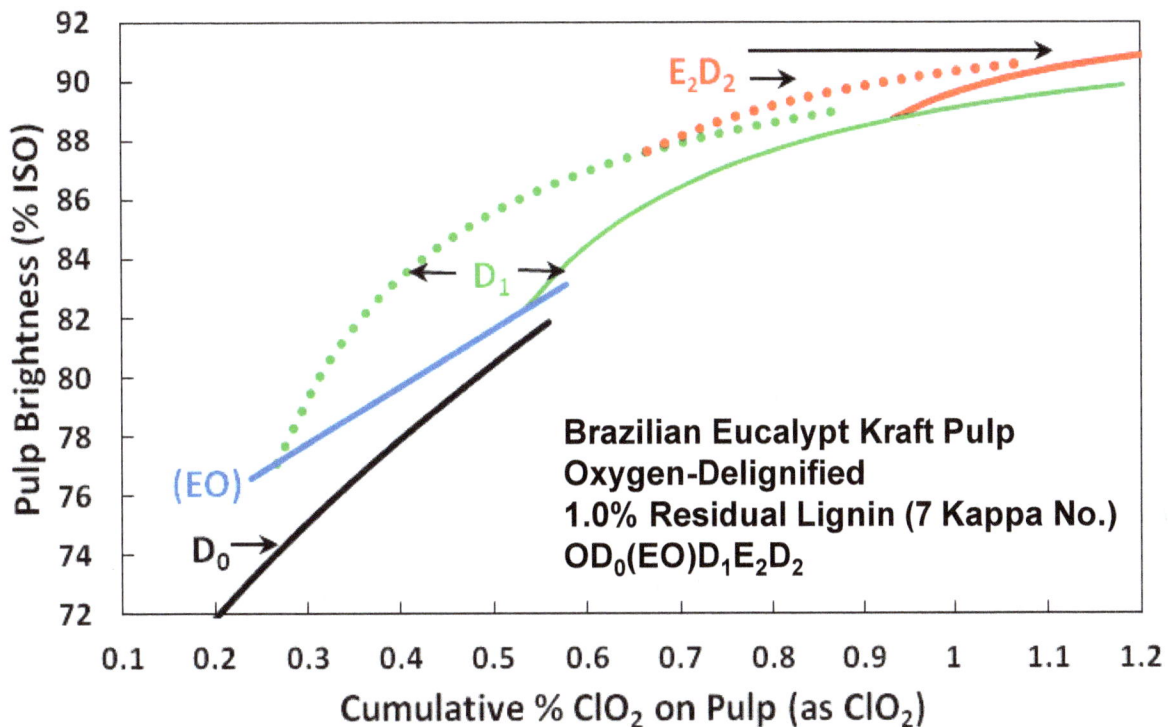

Figure 4.14. Brightening response of a Brazilian oxygen-delignified eucalypt pulp as it progresses through the D(EO)DED sequence as a function of total chlorine dioxide consumed.

Homework

1. What are the three principal components of unbleached kraft pulps? What are their rough percentages in softwood kraft pulps? What are their rough percentages in hardwood kraft pulps?
2. How is residual lignin content in unbleached kraft pulps measured? What is the rough lignin content in unbleached softwood kraft pulps in US? What is the rough lignin content in unbleached hardwood kraft pulps in US?
3. What is pulp brightness? What are the two major methods of measuring pulp brightness?
4. What is pulp viscosity? Why is it important during kraft pulp bleaching? What is the critical limit to TAPPI pulp viscosity for softwoods?
5. What is the approximate limit of delignification that kraft pulping is stopped and bleaching is started for softwoods?
6. What is the symbol used for chlorine dioxide in a bleach sequence? Ozone? Caustic (alkali) extraction? Hydrogen peroxide?
7. What does the beginning of a bleach sequence focus on? What does the latter stages do?
8. What are the four major components of a bleaching stage? What does each component do?
9. What are the three types of bleaching tower designs?

References

The Bleaching of Pulp, 5th Edition, P.W. Hart and A.W. Rudie, Editors; TAPPI Press, Atlanta (2012).

Pulp Bleaching: Principles and Practice, C.W. Dence and D.W. Reeves, Editors; TAPPI Press, Atlanta (1996).

Pulp Bleaching Today, H.U. Suess; Walter de Gruyter GmbH & Co. KG, New York (2010).

2023 TAPPI Bleach Plant Operations Workshop, B.N. Brogdon, Lead Instructor; TAPPI Press, Atlanta (2023).

Chapter 5: Bleaching Stages

I. Introduction

The previous chapter covered general aspects of kraft pulp bleaching as a whole. Pulp bleaching involves multiple stages to first delignify and later brighten the pulp starting at 20% to 45% ISO and going to 84% to 92% ISO. A more detailed account will be provided for the individual delignification and brightening stages in the current chapter.

II. Oxygen Delignification

The purpose of oxygen delignification, sometimes known as oxygen bleaching, is to remove residual lignin from the kraft brownstock. This is done more selectively, both to preserve pulp yield and pulp viscosity or strength, than it is to extend kraft pulping at the digester. Oxygen itself is one of the inexpensive bleaching oxidants, which makes it one of the most economical approaches to remove lignin.

In the early 2020s, oxygen delignification was used in approximately 40% of the bleach plants in the United States, whereas in Canada, it is used in one half of the lines. It can selectively delignify brownstocks between 30% and 45% for hardwood, and between 45% and 60% for softwoods.

How does oxygen delignification accomplish this? It oxidizes the end units (phenolics) of residual lignin. This process converts these groups to more polar organic acids (carboxyl groups), which imparts water solubility/dispersibility in alkali solutions. A benefit of oxygen delignification is its improved environmental performance. It lowers the pulp's kappa number prior to the chlorine dioxide delignification or ozone stage, which lowers operating costs. It reduces the dissolved organic load, such as biological and chemical oxygen demands, that is generated in the effluents. With lower amounts of chlorine dioxide used in bleaching, the amounts of chlorinated organics produced is also lowered. These effluent load reductions are approximately proportional to the level of delignification occurring in the oxygen stage.

The oxygen stage is installed between the brownstock washers and the chlorine dioxide (or ozone) delignification stage. Its effluent can be recycled to be used in brownstock washing. In the late-2010s, the capital cost of installing an oxygen delignification system and associated washers in the United States processing 1,000 metric tons per day is about 33 to 38 million US dollars for a single-stage, or 43 US dollars for a two-stage system.

Oxygen delignification is a heterogeneous reaction that involves a solid (fiber), a liquid (aqueous alkali solution), and a gas (oxygen). It is believed that the delignification kinetics are limited by the mass transfer of oxygen from the gas bubble into the bulk liquid. This mass transport is aggravated by oxygen's low water solubility, which is around 0.8 to 2.8 kg oxygen per metric ton of pulp. Only 10% to 20% of the charged oxygen gas is dissolved in the liquid. Thus, it is important that the gas be finely dispersed within the liquid slurry to promote its dissolution. For a medium consistency system for softwoods, the average dispersed bubble diameter is around 50 to 100 microns, whereas for hardwoods, it is around 200 to 500 microns.

The delignification kinetics are characterized by two distinct phases (Fig. 5.1). The bulk of the initial phase occurs within the first 5 to 20 minutes of the reaction and results in 40% to 70% of the overall delignification. The slow reaction phase happens within the final 30 to 50 minutes.

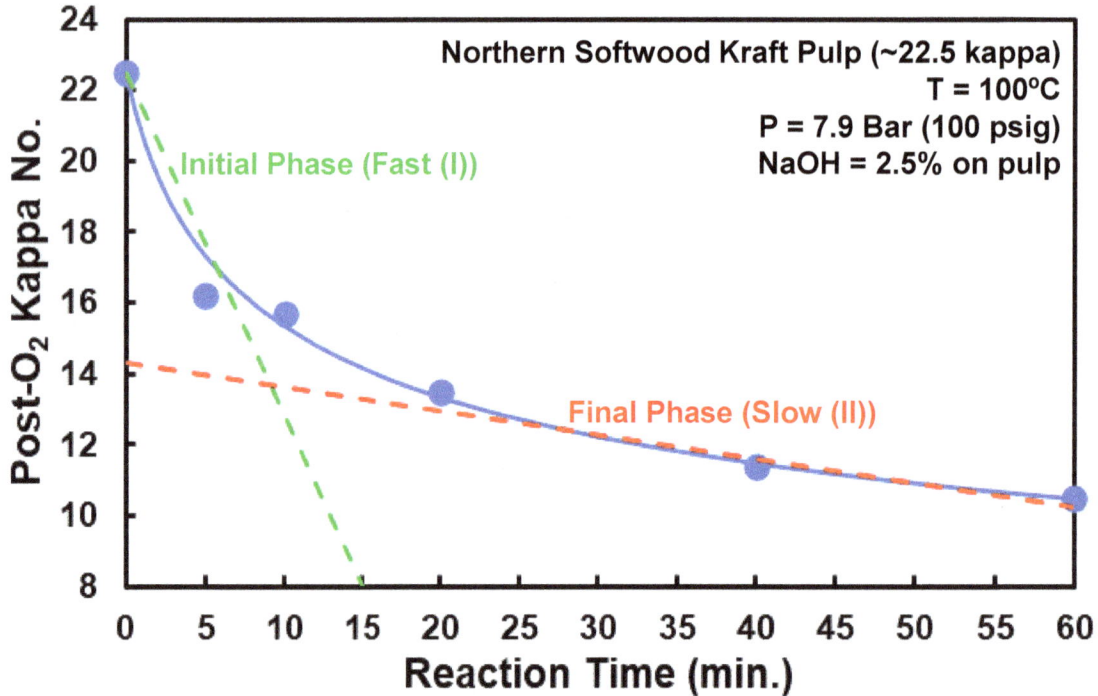

Figure 5.1. Typical medium consistency oxygen delignification kinetics for a softwood brownstock within a single stage.

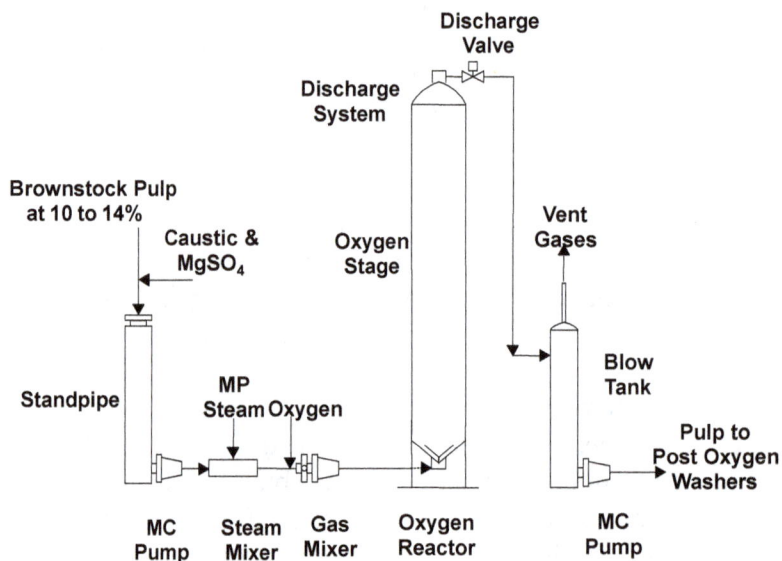

Figure 5.2. Single-stage oxygen delignification system.

A traditional medium-consistency system employs a single pressurized reactor (Fig. 5.2). This design was first introduced in the late 1970s with first-generation high-shear mixers and pumps that could handle the medium consistency stocks. A single-stage system can selectively delignify softwood brownstocks up to 40% to 55% and hardwood brownstocks up to 30% to 40%. This is the most common reactor design used in the United States.

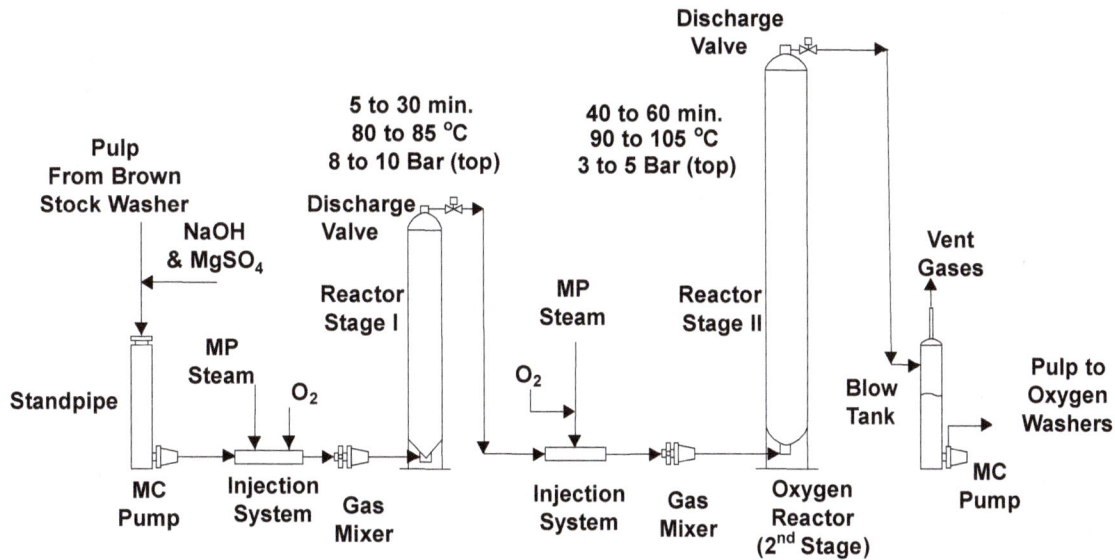

Figure 5.3. Modern two-stage oxygen delignification (O/O) system.

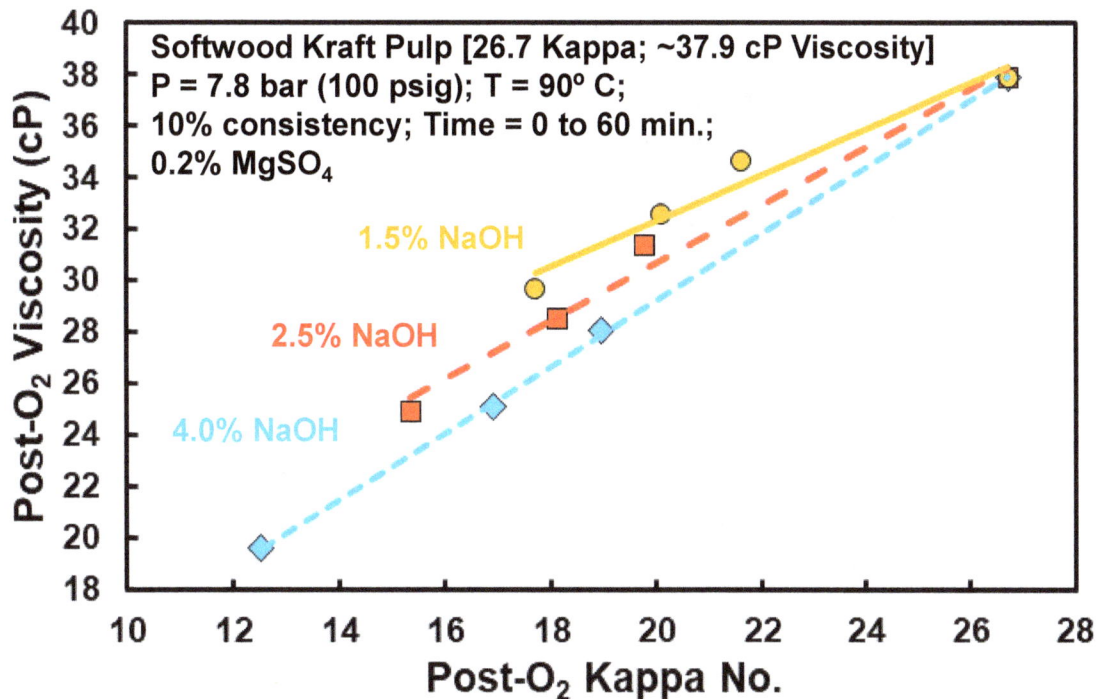

Figure 5.4. Selectivity of the oxygen delignification stage for a northern softwood kraft pulp (26.7 kappa no. and *ca.* 37.9 cP (1175 mL/g)).

71

The newest delignification systems use two pressure vessels, denoted as O/O (Fig. 5.3). The first reactor is smaller than the second reactor. It has a retention time of 10 to 30 minutes and has a higher pressure than the second reactor. The total oxygen and alkali charges used are divided between the two high-shear mixers.

Delignification in an oxygen stage is controlled by the amount of alkali used (Fig. 5.4). This variable affects the viscosity of the pulp. The viscosity lost is proportional to the degree of delignification achieved. From a selectivity standpoint, increasing the alkali level has a mild to negative effect on selectivity. However, a sufficient alkali charge is needed to drive the delignification level to reach 45% to 50%. Typically, the alkali consumption ranges from 0.15% to 0.25% caustic soda on pulp per kappa unit drop. A small amount of magnesium sulfate, approximately 0.2% on pulp, is sometimes used to minimize viscosity losses.

Table 5.1. Typical oxygen delignification conditions used.

Variable	Value
Consistency, %	10 to14
Alkali consumption, kg/t pulp	15 to 35 (1.5% to 3.5% on pulp)
Temperature, °C	85 to 105
Pressure, Bars	Inlet: 7 to 8 / Outlet: 4.5 to 6.0
Retention Time, minutes	50 to 60 (One-Stage); 20/60 (Two-Stage)
Oxygen consumption, kg/t pulp	20 to 24 (2.0 to 2.4%)
MgSO$_4$, kg/t pulp (if needed)	0 – 2.5 (0 to 0.25% on pulp)

Typical operating conditions for single-stage and two-stage oxygen delignification stages are provided in Table 5.1. The alkali used is usually supplied as oxidized white liquor. The effluent from this stage is often used as wash water in brownstock washing. This will add an addition load of 3% to 5% to the causticizing plant and lime kiln in chemical recovery.

III. Chlorine Dioxide Delignification

Chlorine dioxide delignification (D_0) is the most common bleaching stage employed to remove residual lignin from kraft brownstocks and oxygen delignified pulps. In the United States as of the mid-2020s, approximately 60% of bleach plants treat their brownstocks with chlorine dioxide as opposed to oxygen. This stage's function is to remove residual lignin from pulps. It can remove shives and bark (particles), but to a very minor extent. Chlorine dioxide does this by oxidizing and fragmenting the residual lignin to make it less hydrophobic. Some of the lignin is removed in this stage. However, this modified lignin is more susceptible to removal with caustic soda during an extraction stage (E).

The 0 subscript of D_0 indicates that this chlorine dioxide stage occurs prior to brightening stages, such as D_1 or D_2. It designates that the chlorine dioxide is being used for kappa number reduction. Another subscript used in the older bleaching literature is 100 (D_{100}). It notes that 100% chlorine dioxide substitution is used in a chlorination stage. It was used in the 1980s and 1990s when partial chlorine dioxide substitution in the chlorination stage was common.

The chlorine dioxide used in the D_0 stage is often expressed as kappa factor (KF). It is sometimes called active chlorine multiple (ACM) or equivalent chlorine multiple. It is the ratio of the percentage chlorine dioxide used on pulp, expressed as active chlorine, to the kappa number of the pulp entering the D_0 stage:

$$Kappa\ Factor\ (KF) = \frac{\%\ active\ Cl_2\ on\ pulp}{Kappa\ number} = \frac{\%\ ClO_2\ on\ pulp\ (as\ ClO_2)\ x\ 2.63}{Kappa\ number}$$

Residual lignin is oxidized by chlorine dioxide under acidic conditions (2.5 to 3.5 pH). It yields a partially modified lignin that is smaller in size. The phenolic end groups are oxidized to organic acids and to quinones. This makes the modified lignin partially soluble in the acid bleaching liquor and more soluble in the alkaline extraction stage, where the acids are ionized. Chlorine dioxide is selective, preferring to react with residual lignin rather than with the carbohydrates. This results in minimal carbohydrate degradation and pulp viscosity losses. Some key variables for a D_0 stage are given in Table 5.2. For brevity, only the first three items will be covered.

Table 5.2. Chlorine dioxide delignification (D_0) variables.

Softwood or Hardwood Pulp
Initial Kappa Number
ClO$_2$ Chemical Charge (Kappa Factor)
Reaction Temperature & Time
Reaction pH
Consistency
Incoming Washer Carryover

Figure 5.5 shows how the chlorine dioxide charge in D_0 affects the kappa number of a United States southern pine brownstock. The D_0 kappa number decreases as the kappa factor charge is increased in a nearly linear fashion (blue solid line). The slope for softwoods is approximately 6.4 kappa units per 1% chlorine dioxide (as ClO_2) consumed. Only a part of the oxidized lignin is removed. The full extent of the D_0 stage on delignification is not seen until after the extraction stage (dotted lines). It is important to note that if the extraction stage uses oxidant reinforcement, this too affects the kappa number drop. This will be discussed later.

73

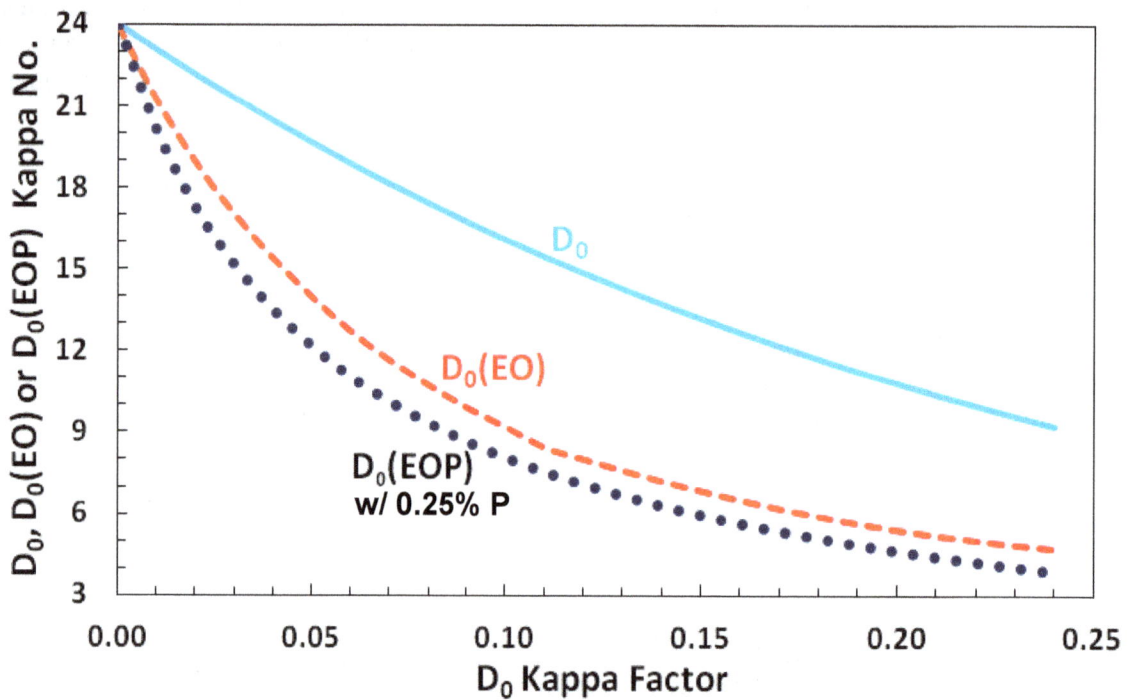

Figure 5.5. Kappa number reductions in the D_0, $D_0(EO)$, and $D_0(EOP)$ *versus* D_0 kappa factor used for a United States southern pine brownstock (initial kappa number of 24).

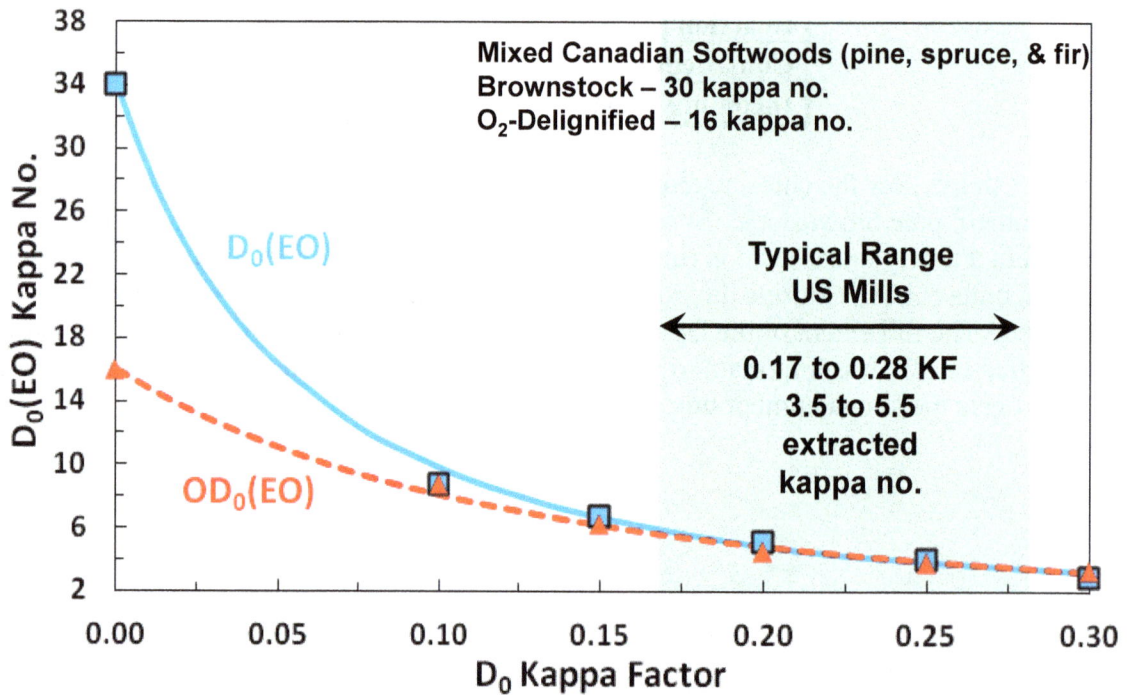

Figure 5.6. Kappa number reductions in the $D_0(EO)$ *versus* D_0 kappa factor used for a Canadian kraft brownstock (initial kappa number of 30) and oxygen delignified pulp (initial kappa number of 16).

Figure 5.6 illustrates how the kappa factor charge affects the delignification for a Canadian brownstock and an oxygen-delignified pulp. The kappa drop for the oxygen delignified pulp is smaller than for the brownstock at a given kappa factor charge. However, it should be noted that the actual amount of chlorine dioxide used is approximately 50% lower. Final extracted kappa numbers at a kappa factor charge greater than 0.15 are similar for both pulps. The highlighted area is the typical kappa factor charges used in the United States for softwoods.

Figure 5.7 demonstrates the effect of kappa factor charges on $D_0(EO)$ kappa number for a United States hardwood kraft pulp. A difference between hardwood and softwood pulps is that the extracted kappa continues to decrease linearly from 0.10 to 0.30 kappa factor. This trend levels-off around 0.25 to 0.30 kappa factor for softwoods. The slope for hardwoods in D_0 stage alone is roughly 4.3 to 6.1 kappa units per 1% chlorine dioxide (as ClO_2) consumed. The highlighted area is the common kappa factor charges used in the United States for mixed hardwood pulps.

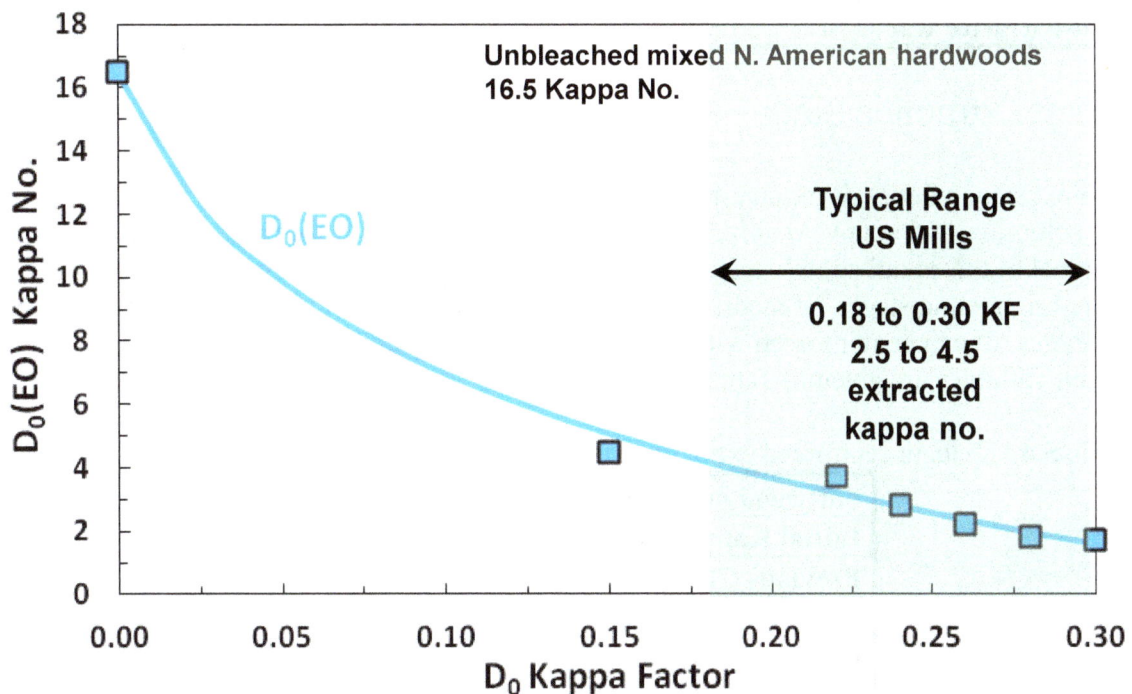

Figure 5.7. Kappa number reductions in the $D_0(EO)$ *versus* D_0 kappa factor used for a United States mixed hardwood kraft pulp (initial kappa number of 16.5).

Listed in Table 5.3 are common D_0 stage conditions for softwood and hardwood brownstocks in United States bleach plants. The D_0 stage is typically run at a pH of 2.5 to 3.5 to fully use the oxidation power of chlorine dioxide and its intermediate species. This pH range also minimizes calcium oxalate and barium sulfate scales. Brownstock washing is not perfect, so it contributes to delignification inefficiencies. General levels are on the order of 10 kg saltcake/t pulp and will consume approximately 0.03 to 0.07 of the kappa factor charge. The D_0 stage temperature typically runs from 45°C to 65°C for a reaction time of 30 to 60 minutes. The total capital cost of installing chlorine dioxide delignification system utilizing 1000 metric tons per day is approximately 19 US dollars (as of late 2010s).

Table 5.3. Typical D_0 stage conditions for softwood (SW) and hardwood (HW) pulps in United States bleach plants in the mid-2010s. (LC is low consistency and MC is medium consistency.)

Variable	Brownstock		Oxygen-Delignified	
	SW	HW	SW	HW
Kappa No. to D_0	24 – 30	14 – 16	13 – 16	8.5 – 12.5
D_0 Kappa Factor	0.18 – 0.25	0.24 - 0.37	0.23 – 0.37	0.21 – 0.39
D_0 Consistency, %	3 – 4 LC 10 – 11 MC	3.5 – 4.0 LC 9.5 – 11 MC	3 – 4 LC 9.0 – 11.5 MC	3.5 – 4.0 LC 10.0 – 10.5 MC
H_2SO_4, % on pulp	0 – 1.5	0 – 1.6	0 – 1.5	0 – 2.2
Vat pH	2.2 – 2.8	2.5 – 2.8	3.2 – 2.9	2.4 – 3.1
Temperature, °C	45 – 60	50 – 60	55 – 75	55 – 75
Time, minutes	30 – 60	30 – 60	30 – 60	30 – 60
Carryover, kg Na_2SO_4/t pulp	5 – 30	12 – 20	3 – 15	7 – 12
Post-Extracted Kappa No.	3.5 – 5.5	2.5 – 4.5	2.0 – 4.0	2.0 – 5.0

II. First Extraction Stage

The next stage after D_0 is oxidant-reinforced extraction (*i.e.*, (EO), (EOP) or (EP)). Its purpose is to solubilize and remove oxidized lignin that was not removed from an acid oxidation stage. Removal of this alkali-soluble material helps to save on oxidant chemical consumption used in future brightening stages. It modifies the oxidized lignin so that it will react again with bleaching agents. It can be reinforced with small doses of oxygen and/or peroxide. Some of the important variables are given in Table 5.4.

Table 5.4. Oxidant-reinforced extraction stage ((EO) or (EOP)) variables.

Softwood or Hardwood Pulp
Initial Kappa Number
Previous ClO_2 Chemical Charge (Kappa Factor)
Reaction Temperature & Time
Caustic Charge & pH
Oxygen and/or Peroxide Charge
Consistency
Incoming Washer Carryover

Simple extraction with caustic alone (E_1) results in a 3 to 5 kappa drop for a D_0 treated pulp (Fig. 5.8). Reinforcement of this stage with oxygen (*i.e.*, (EO)) can enhance this delignification by an additional 1 to 2 units at a given chlorine dioxide charge in the previous D_0 stage. Adding peroxide to (EO) stage can boost the delignification even further by an additional 0.5 to 1 units. Peroxide reinforcement in an (EO) can also lower the amount of chlorine dioxide used in the D_0 to reach a target extracted kappa.

The extraction stage after D_0-treatment for low-kappa oxygen delignified pulps is less dramatic (Fig. 5.9). Oxygen reinforcement provides a minor change in extracted kappa. Some additional delignification is observed when (EOP) stage is used versus E_1 or (EO). Thus, caution should be exercised when generalizing extraction stage conditions for softwoods *versus* hardwoods.

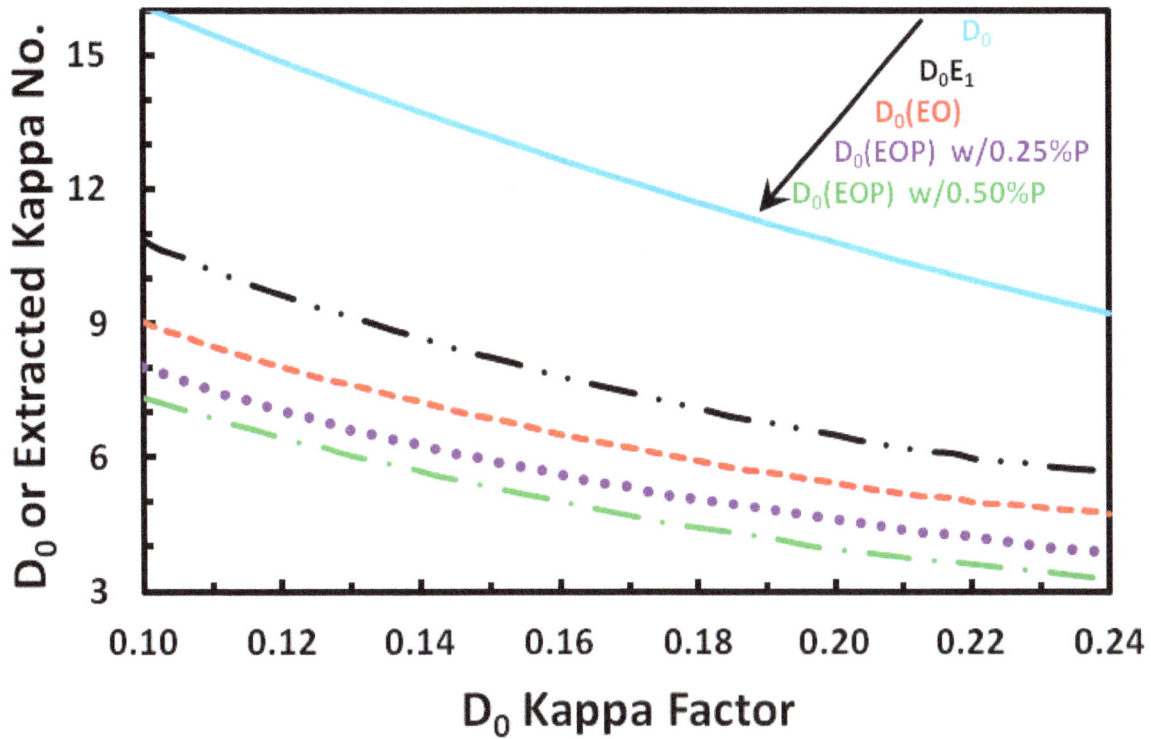

Figure 5.8. Kappa number reductions in various extraction stages *versus* D_0 kappa factor used for a United States southern pine brownstock (unbleached kappa number of 24).

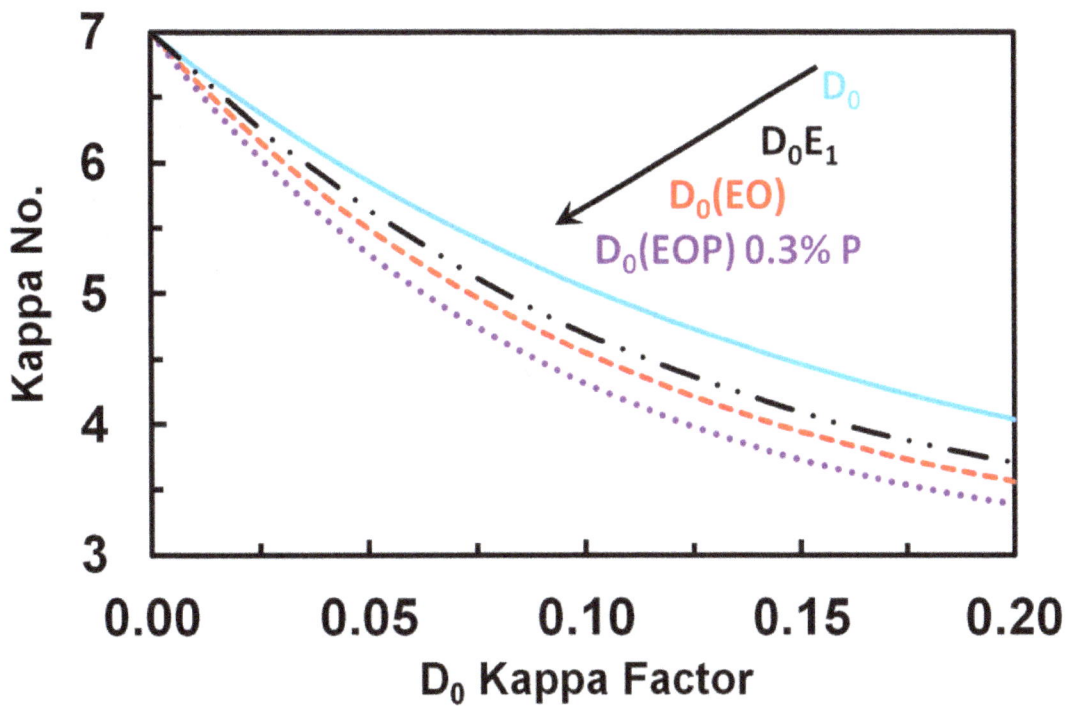

Figure 5.9. Kappa number reductions in various extraction stages *versus* D_0 kappa factor used for a Brazilian oxygen delignified eucalypt pulp (oxygen-delignified kappa number of 7).

Oxygen reactions in extraction occur quickly, within the first few minutes after it is mixed with the slurry (Fig. 5.10). Sufficient pressure, either supplied by the hydrostatic head in the pre-retention tube, or over pressurization of the reactor for the first 10 minutes, achieves the full benefits of oxygen reinforcement. Generally, 2.0 to 3.1 bar (30 to 45 psig) of oxygen pressure is sufficient. The oxygen consumed by the pulp is roughly 0.2 to 0.4% on pulp. It is important to note that the oxidized lignin after an (EO) stage is nearly the same as that contained after an E_1 even though the (EO) kappa is a few units lower. This reflects that the remaining lignin is more oxidized after an (EO) than after an E_1.

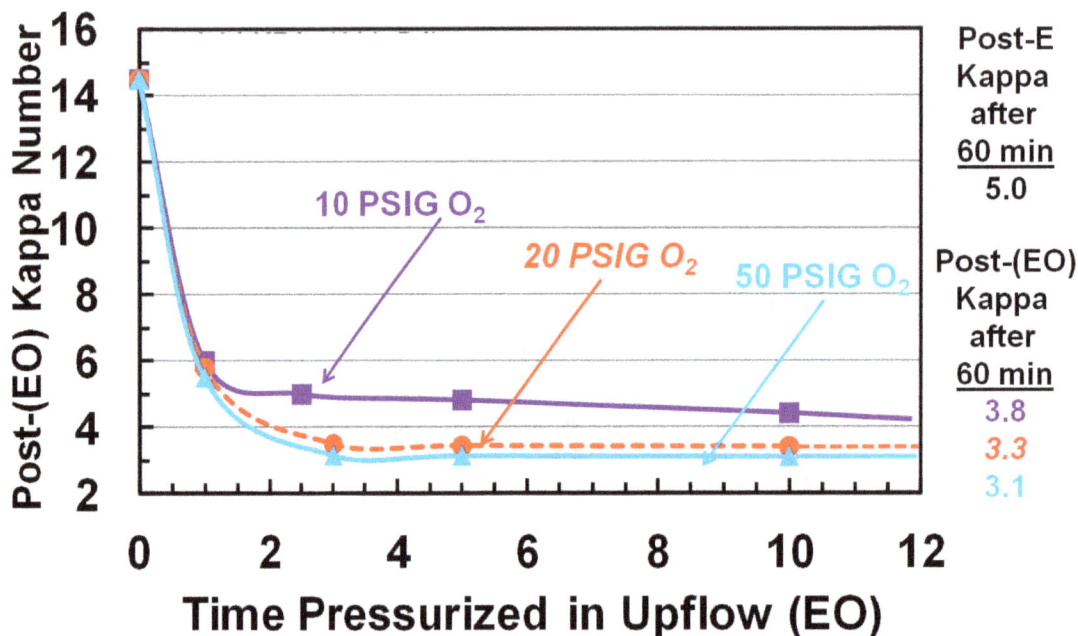

Figure 5.10. Effect of oxygen pressure (10, 20, and 50 psig (0.7, 1.4, and 3.5 bar)) and time (minutes) in an (EO) stage *versus* the post-extracted kappa number of a Canadian softwood kraft pulp (pre-(EO) kappa number of 14.5).

Adding 0.25% to 0.50% peroxide to an (EO) stage increases pulp brightness by 5 to 10 points, as is shown in Figure 5.11. It also decreases the extracted kappa of the pulp by 0.4 to 0.8 kappa units (Fig. 5.12). These brightness gains do not fully carry forward into future brightening stages, but they can influence bleach usage in later stages if the gains are greater than 5 points. Increasing the temperature of an (EOP) by $\Delta 10°C$ ($\Delta 18°F$) increases the kappa number drop by 0.25 units and increases brightness by 2.5 points. The lower kappa of an (EOP) *versus* an (EO) reflects more oxidized lignin being removed. Most all the peroxide reacts with oxidized lignin in pulp, not dissolved lignin in effluent.

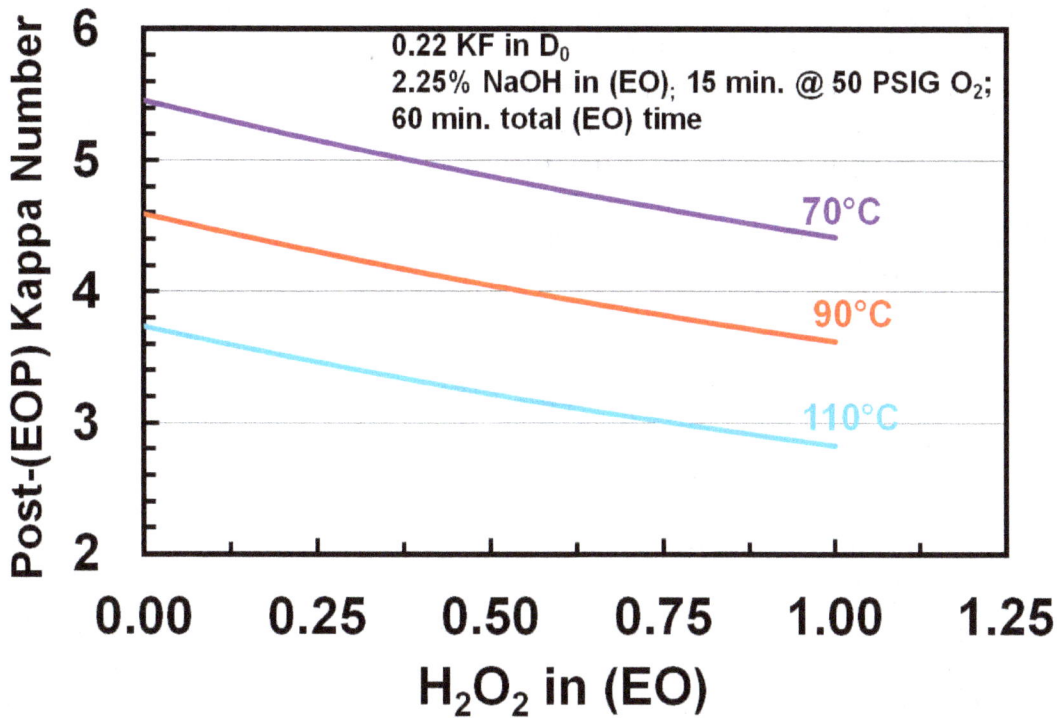

Figure 5.11. Effect of peroxide addition and temperature for an (EOP) stage on the post-extracted kappa number of a Canadian softwood kraft pulp (brownstock kappa no. 31.4).

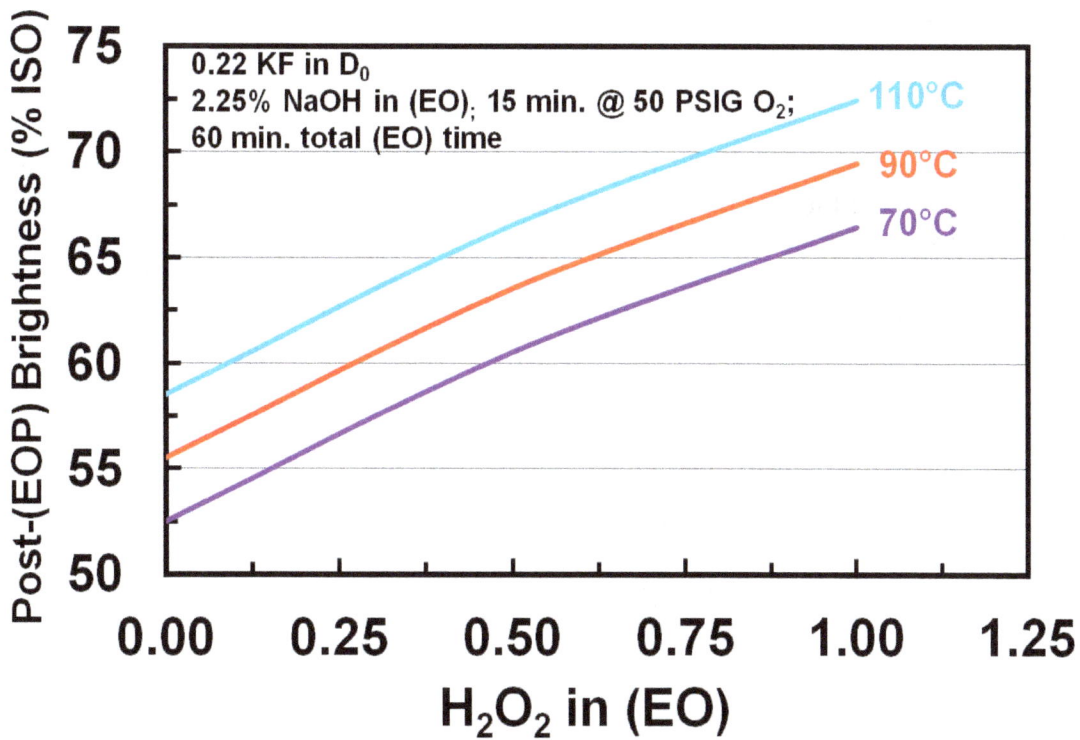

Figure 5.12. Effect of peroxide addition and temperature for an (EOP) stage on the post-extracted brightness of a Canadian softwood kraft pulp (brownstock kappa no. 31.4).

Recorded in Table 5.5 are common oxidant reinforced extraction conditions used for various pulps in the United States. The amount of caustic used, as a percentage on pulp, is roughly the entering brownstock kappa number divided by ten. This corresponds to pH values that range from 10.4 to 11.4. The amount of oxidant reinforcement is usually low, less than 1% on pulp, with somewhat more oxygen used than peroxide. Temperatures typically used range from 76°C to 84°C, and reaction times usually run from 45 to 120 minutes. The total capital cost of installing oxidant-reinforced extraction system utilizing 1000 metric tons per day is approximately 11 to 15 US dollars (as of late 2010s).

Table 5.5. Typical oxidant reinforced extraction conditions for softwood (SW) and hardwood (HW) pulps in United States bleach plants in the mid-2010s.

Variable	Brownstock		Oxygen-Delignified	
	SW	HW	SW	HW
Prior to Extraction				
Kappa No. to D_0	24 – 30	14 – 16	13 – 16	8.5 – 12.5
D_0 Kappa Factor	0.18 – 0.25	0.24 - 0.37	0.23 – 0.37	0.21 – 0.39
Extraction Stage				
NaOH, % on pulp	1.7 – 2.6	1.0 – 1.9	1.5 – 2.3	0.9 – 1.5
O_2, % on pulp (EO) or (EOP)	0.50 – 0.75	0.40 – 0.55	0.40 – 0.55	0.25 – 0.50
H_2O_2, % on pulp (EOP)	0.30 – 0.70	0.20 – 0.55	0.25 – 0.40	0.10 – 0.40
Extracted Kappa No.	3.5 – 5.5	2.5 – 4.5	2.0 – 4.0	2.0 – 5.0
Vat pH	10.2 – 10.9	10.3 – 11.4	10.5 – 11.1	10.3 – 11.3
Temperature, °C	77 – 84	76 – 82	76 – 84	70 – 80
Time, minutes	45 - 120	45 - 120	45 - 120	45 - 120

V. First Chlorine Dioxide Brightening Stage

After leaving the first extraction stage, the pulp switches from being delignified to being brightened. Chlorine dioxide used for brightening is designated by D_1. Here, chlorine dioxide removes residual chromophores, and in certain circumstances, shives and dirt. The pulp entering this stage will contain traces of lignin of between 2.5 and 5.5 kappa (0.45% and 1.0% oxidized lignin). The D_1 stage will remove the traces of lignin to non-detectable levels (< 2.5 kappa). Table 5.6 lists some of the main variables of the D_1 stage. The first three variables will be covered.

Table 5.6. First chlorine dioxide brightening (D_1) stage variables.

Softwood or Hardwood Pulp
Extracted Kappa Number
ClO_2 Chemical Charge
Reaction Temperature & Time
Reaction pH
Consistency
Residual ClO_2
Extraction Washer Carryover

The chlorine dioxide needed for brightening depends on the post-extraction kappa entering the D_1 stage and the end brightness target (Fig. 5.13). This relationship is influenced by the wood species but appears independent of prior bleaching delignification treatment. Typically, the optimum extracted kappa is between 3.0 and 4.5 for softwoods. To reach a brightness of 85% ISO, a pulp with an extracted kappa of 3.2 will require 0.9% chlorine dioxide on pulp. If the extracted kappa were to increase to 4.2, then the amount of additional bleach would increase by 67% to 1.5% chlorine dioxide on pulp.

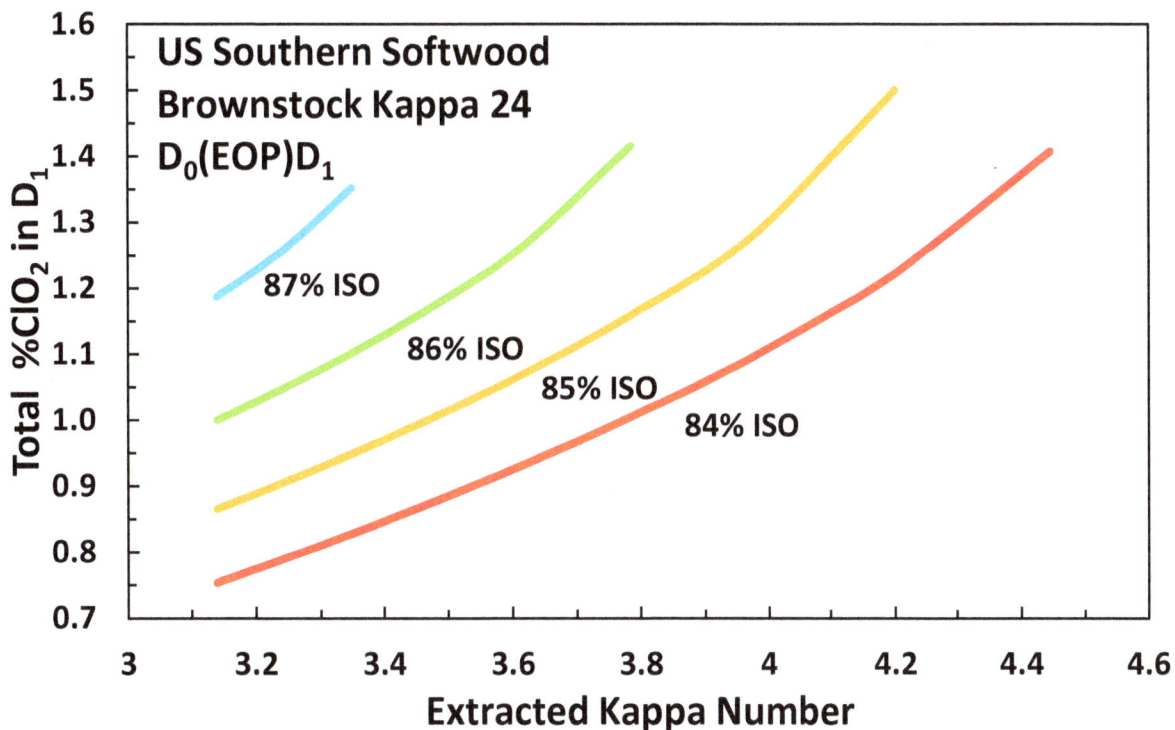

Figure 5.13. Effect of extracted kappa number on total chlorine dioxide used in the D_1 for brightness targets between 84% and 87% ISO for a United States southern softwood brownstock treated with the $D_0(EOP)D_1$ sequence.

Figure 5.14 illustrates the chlorine dioxide needed to achieve an 86% ISO brightness for sweetgum, maple and oak. Most mills do not use a single hardwood specie, so this graph shows the effect of species variation that may be seen when the chip supply is varied. For example, with a post (EO) kappa of 4, the amount of chlorine dioxide required when swinging from oxygen delignified oak to a brownstock sweetgum stock will increase from 0.3% to 0.8% on pulp.

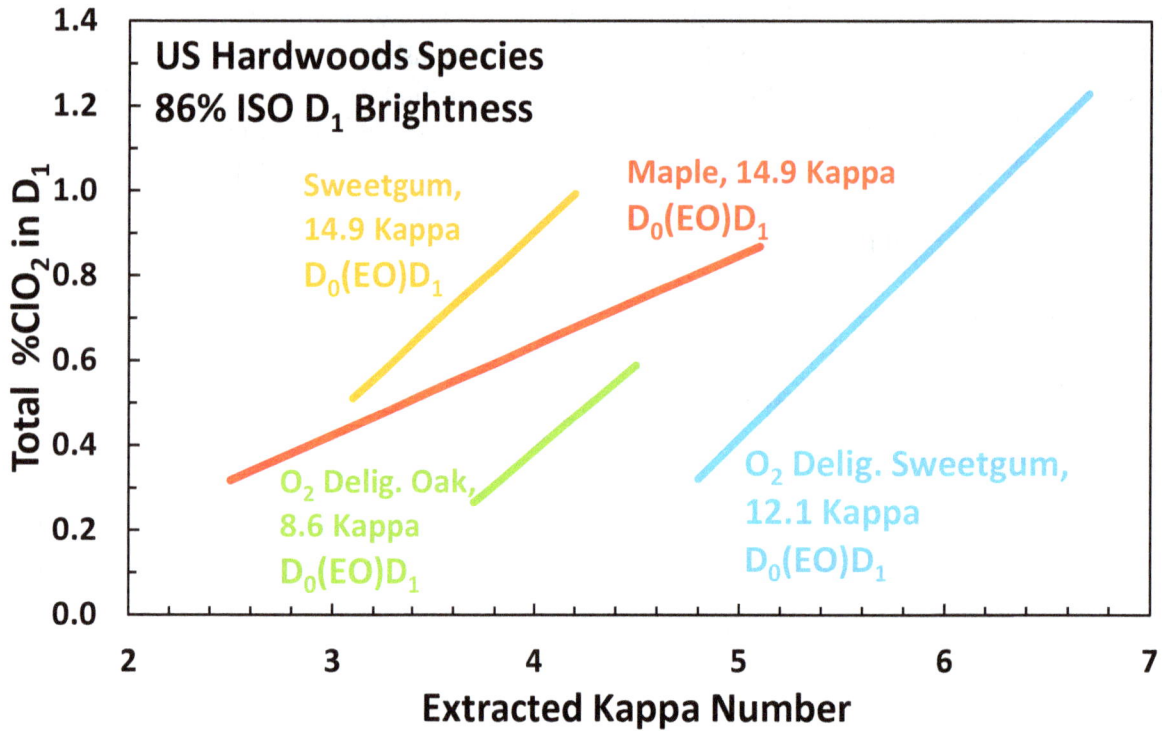

Figure 5.14. Effect of extracted kappa number on total chlorine dioxide used in the D_1 for brightness target of 86% ISO for various individual North American hardwood species treated with the $D_0(EO)D_1$ or $OD_0(EO)D_1$ sequence.

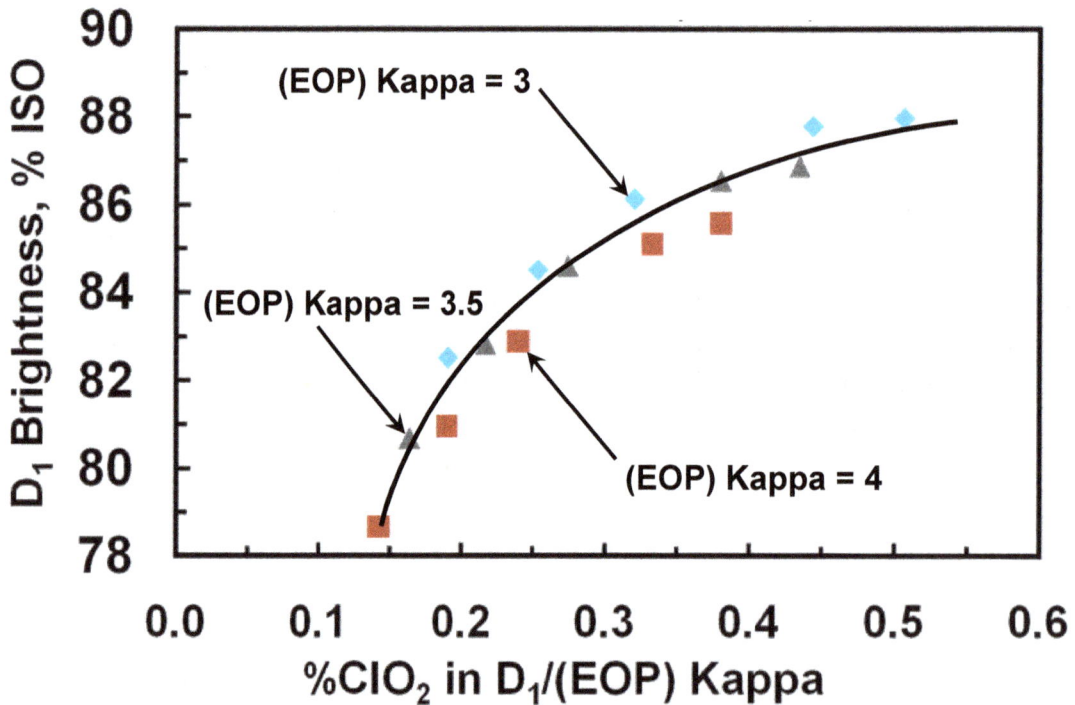

Figure 5.15. Brightening response in the D_1 stage *versus* of chlorine dioxide multiple (% ClO_2/((EOP) kappa no.)) used for a United States southern pine pulp bleached by $D_0(EOP)D_1$.

The amount of chlorine dioxide used in the D_1 depends on the brightness target and the kappa number entering the stage. Typically, 0.2 to 0.3% ClO_2 on pulp is used for each kappa unit entering the D_1 stage (Fig. 5.15). Mills may use more than this, but the brightening efficiency of chlorine dioxide drops off significantly. Extraction washer carryover from imperfect washing can add an additional 0.5 to 2.0 kappa load to the D_1 stage, which can contribute to D_1 brightening inefficiencies. This results in 0.2% to 1.0% extra chlorine dioxide (as ClO_2) per ton of pulp being consumed.

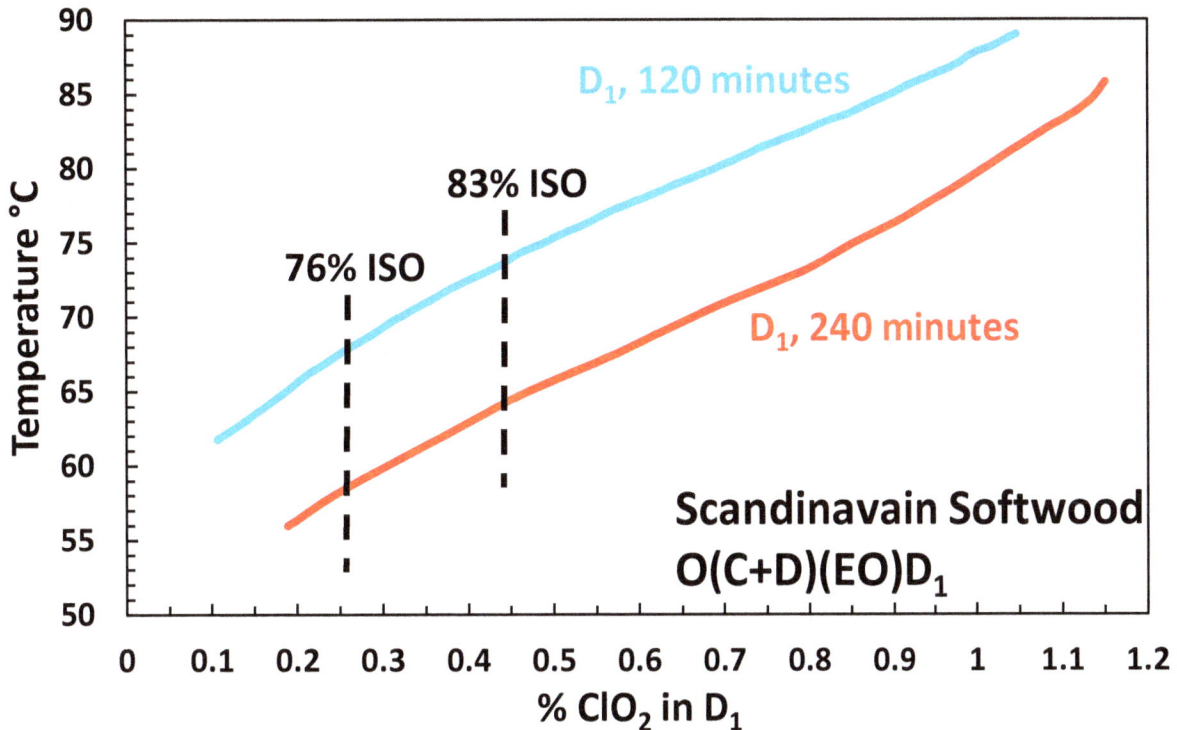

Figure 5.16. Temperature in D_1 stage to reach trace residual chlorine dioxide (0.04 g/L) for a Scandinavian softwood bleached with $O(C+D)(EO)D_1$ sequence.

The usual conditions applied in the D_1 are listed in Table 5.7 for United States softwood and hardwood pulps. The optimum terminal pH is around 4 for brightening but varies mill to mill depending on whether it is a softwood or hardwood mix. Caustic or sulfuric acid may be added to adjust the pH. Mills may run lower pH (*i.e.*, 3.0) to bleach shives and bark. This makes the stage behave more like a delignification stage and lowers the brightness that can be obtained. The D_1 stage operates with a small residual (≤ 0.05 g ClO_2/L) to prevent thermal brightness reversion if the bleach is totally exhausted (Fig. 5.16). The reaction time runs from 120 to 180 minutes at temperatures between 70°C and 80°C. In various designs, a 20-to-30-minute pre-retention upflow tube is used followed by 100 to 160 minutes downflow tower. This allows the fast-brightening reactions to happen with the higher bleach concentrations, while the subsequent downflow tower allows for slower reactions to complete at the lower bleach levels. Bleach consistencies of 9% to 16% do not have a major effect on brightening so long as tower channeling does not happen.

Table 5.7. Typical first chlorine dioxide brightening conditions for extracted softwood and hardwood pulps in United States bleach plants in the mid-2010s.

Variable	Value
Prior to D_1	
Extracted Kappa No.	$2.6 - 5.1$
Extraction Carryover	*ca* 0.50 kappa units (or 0.44 kg COD/t pulp)
D_1 Stage	
ClO_2 Multiple, % ClO_2/Kappa	$0.20 - 0.35$
ClO_2, % on pulp	$0.60 - 1.20$
NaOH or H_2SO_4, % on pulp	$0 - 0.20$ NaOH (for $0.7\% - 1.3\%$ ClO_2) $0 - 0.10$ H_2SO_4 (for $0.7\% - 0.5\%$ ClO_2)
ClO_2 residual, g/L.	trace $- 0.05$
Brightness, % ISO	$82 - 88$
Vat pH	$3.0 - 4.0$
Temperature, °C	$70 - 80$
Time, minutes	$120 - 180$

VI. Second Extraction and/or Second Chlorine Dioxide Brightening Stage

The first chlorine dioxide brightening stage will brighten a softwood kraft pulp up to 88% ISO brightness, and a mixed hardwood pulp up to 89% ISO. To reach 90% to 92% ISO brightness, additional bleaching is required. This is most often accomplished by employing a second chlorine dioxide brightening stage (D_2), either with or without a preceding second extraction stage (E_2). Addition of a D_2 helps to obtain 89% to 92% ISO by lowering bleach usage by redistributing the chlorine dioxide spent for both delignification and brightening.

Some bleach sequences use an E_2 stage, which is placed between the first and second chlorine dioxide brightening stages. Its function differs completely from that of the (EO) or (EOP). This stage does not remove lignin or chromophores. An E_2 stage treatment does not increase the brightness of the D_1 pulp to a significant extent. Instead, the E_2 reactivates the pulp to make it easier to react with chlorine dioxide again. This is shown in Figure 5.17. Also shown is an abbreviated E_2 that is designated as N in D_N or as (D_1/E_2). Here, the D_1 pH at the end is readjusted from 4 to 10 for 1 to 5 minutes and then subjected to washing and to D_2 bleaching. The brightening effect of such a treatment lies between that of D_1D_2 and $D_1E_2D_2$.

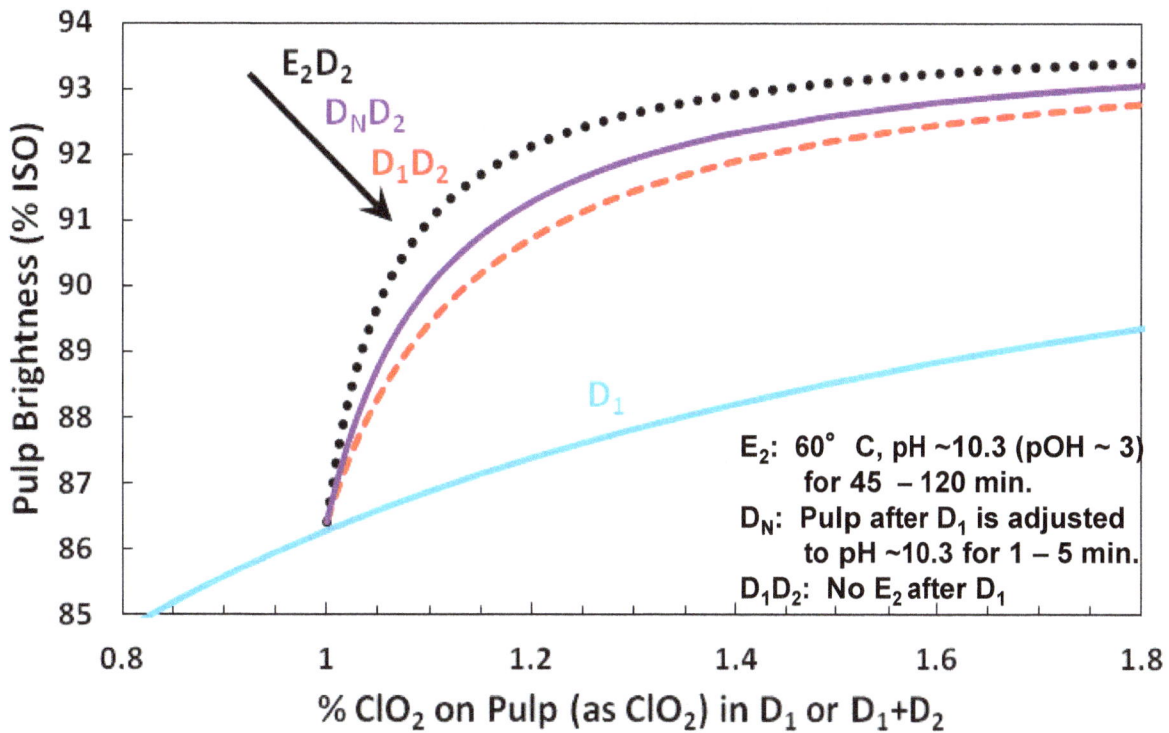

Figure 5.17. Brightness of an O(C+D)E$_1$ Scandinavian softwood kraft pulp treated by D$_1$D$_2$, D$_N$D$_2$, (or (D$_1$/E$_2$)) and D$_1$E$_2$D$_2$ sequences. D$_N$ is like D$_1$ except that the pulp's pH is adjusted at the end from 4 to 10.3 before washing.

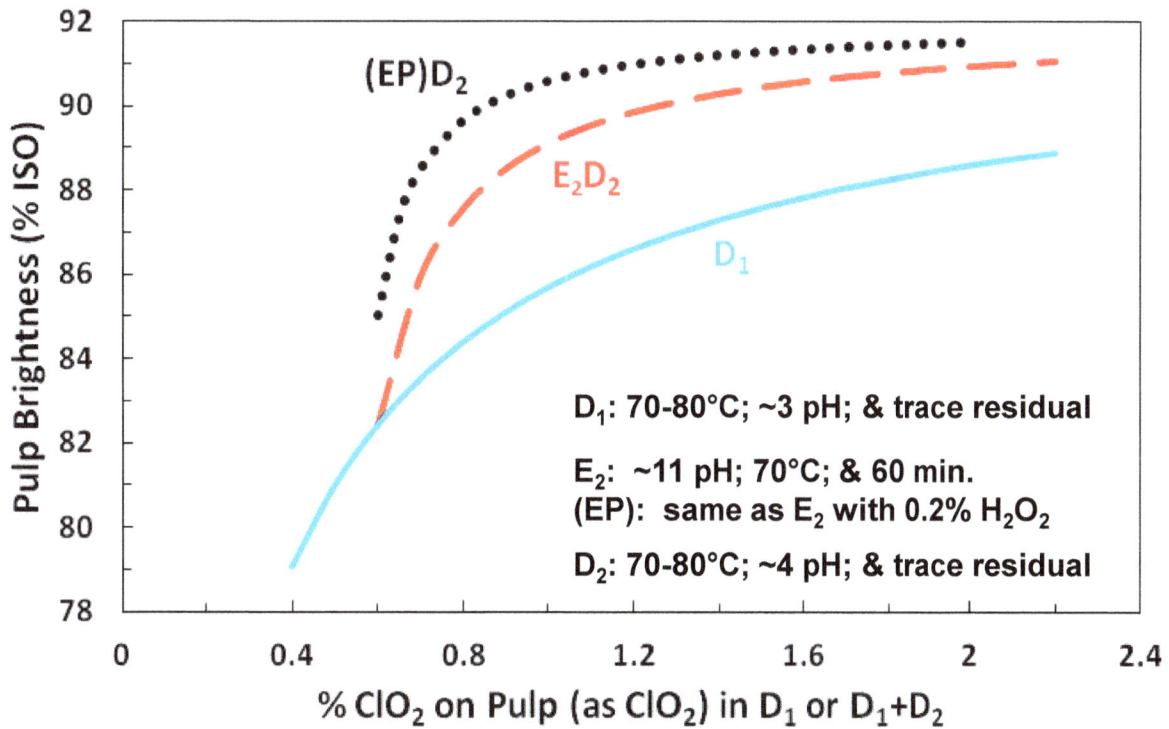

Figure 5.18. Brightness of a softwood kraft pulp treated by D$_0$(EO)D$_1$E$_2$D$_2$ and D$_0$(EO)D$_1$(EP)D$_2$ sequences.

Some E_2 stages are reinforced with low levels of peroxide; they are designated as an (EP) stage (Fig. 5.18). Adding 0.2% peroxide to this stage can replace 0.4% to 0.6% chlorine dioxide needed to reach high brightness. Oxygen reinforcement is not used in the E_2 position since it does not augment brightness development. Table 5.8 displays some of the common conditions used during an E_2 or (EP) stage. Such a stage is conducted at medium pulp consistency (10% to 14%).

Table 5.8. Typical second extraction stage conditions used in United States bleach plants in the mid-2010s.

Variable	E_2 or (EP)		N in D_N	
	SW	HW	SW	HW
NaOH, % on pulp	0.2 – 0.8	0.4 – 1.0	0.5 – 0.9	0.7 – 0.8
H_2O_2, % on pulp for (EP)	0.1 – 0.5	0.05 – 0.2	None	None
Vat pH	9.5 – 10.6	8.5 – 11.0	9.5 – 10.5	8.5
Temperature, °C	60 – 80	70 – 80	70 – 75	70 – 80
Time, minutes	40 – 120	35 – 110	3 – 5	3 – 5

A second chlorine dioxide brightening stage is used to obtain pulp brightness levels of 89 to 92% ISO. Although the D_2 stage will continue to minimize bleachable dirt, normally the D_1 stage will have eliminated the bulk of it, so that the D_2 stage can focus on brightness development. The D_2 is done either after the D_1, E_2 or (EP) stage. Typically, the chlorine dioxide charge is low, 0.2% to 0.5% on pulp, since it reacts sluggishly with the residual chromophores left in the pulp.

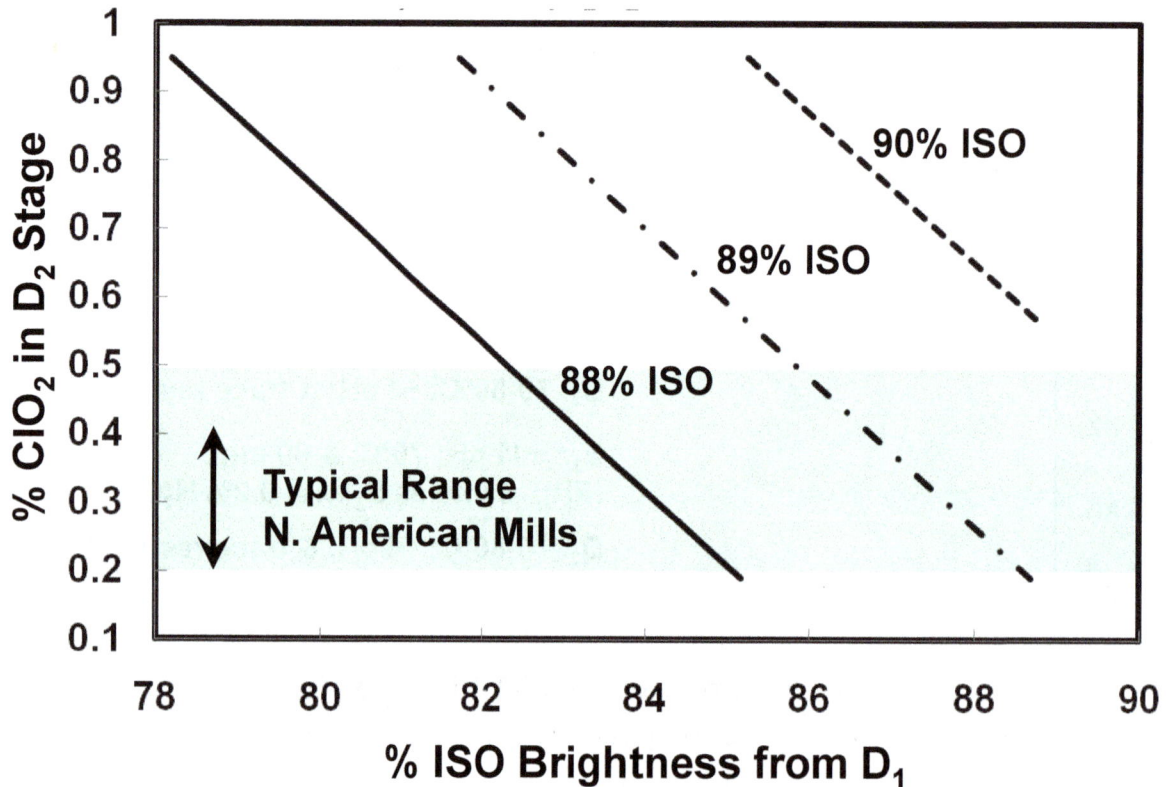

Figure 5.19. Brightening response of a United States southern pine pulp treated by the $D_0(EOP)D_1E_2D_2$ sequence.

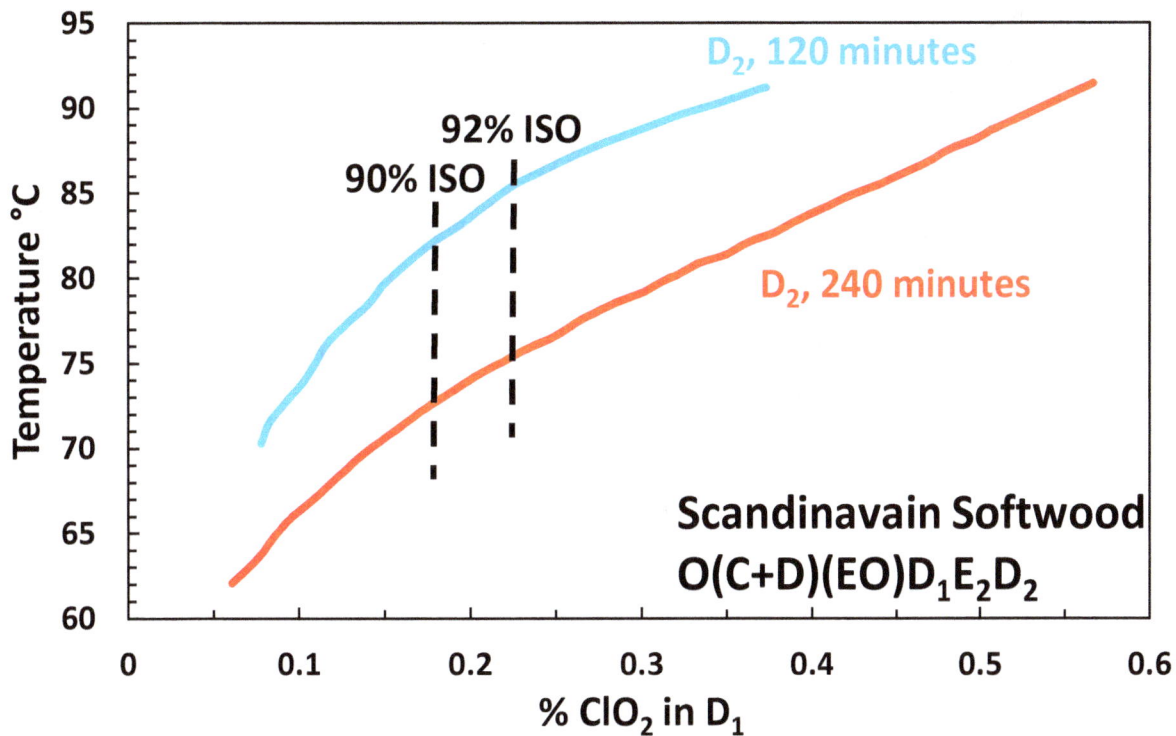

Figure 5.20. Temperature in D_2 stage to reach trace residual (0.04 g ClO_2/L) for a Scandinavian softwood bleached by $O(C+D)(EO)D_1E_2D_2$ sequence.

The chlorine dioxide needed to reach a D_2 brightness is related to the D_1 exit brightness, as is shown in Figure 5.19. It should be observed that this linearity exists only over a small range. The highlighted area represents the typical dosages applied in United States mills. Similar trends are noted for other softwoods and hardwoods but will not be quite the same.

Time and temperature are related. While a fixed time limits the overall reaction, temperature will increase the D_2 reaction rate. Figure 5.20 illustrates this at 2- and 4-hour retention times for a D_2 stage for a Scandinavian softwood. Therefore, higher chlorine dioxide charges require a higher tower temperature and/or longer retention time. A shorter retention time can be compensated for by increasing tower temperature, and *vice versa*.

Table 5.9 presents common conditions employed in D_2 stages in the United States. The brightness entering the D_2 stage is usually between 80% ISO and 87% ISO. A bleaching temperature between 70°C and 80°C and a reaction time between 150 and 200 minutes are used. The terminal D_2 pH of 3.6 to 4.5 is often observed, which is higher than that observed for the D_1. Chlorine dioxide residuals run from 0.01 to 0.07 g/L. These results for a $D_0(EO)D_1E_2D_2$ sequence require a softwood extracted kappa number between 4.0 and 5.0 *versus* a kappa between 3.3 and 4.0 for the shortened $D_0(EO)D_1$ sequence.

Table 5.9. Typical second chlorine dioxide brightening stage conditions used for softwoods and hardwoods in United States bleach plants in the mid-2010s.

Variable	Value
ClO_2, % on pulp	$0.2 - 0.5$
NaOH or H_2SO_4, % on pulp	variable
ClO_2 residual, g/L.	$0.01 - 0.07$
Brightness, % ISO	$88 - 90$
Vat pH	$3.6 - 4.5$
Temperature, °C	$70 - 80$
Time, minutes	$150 - 200$

VII. Other Bleaching Stages

VII.1. Hydrogen Peroxide Brightening

A select few mills in the United States end their bleaching sequences with a peroxide stage. In these cases, the terminal peroxide stage is preceded by a first chlorine dioxide brightening stage. The amount of peroxide used is somewhat higher than that used in an (EP) stage, typically between 0.3% to 1.0% on softwood and hardwood pulps.

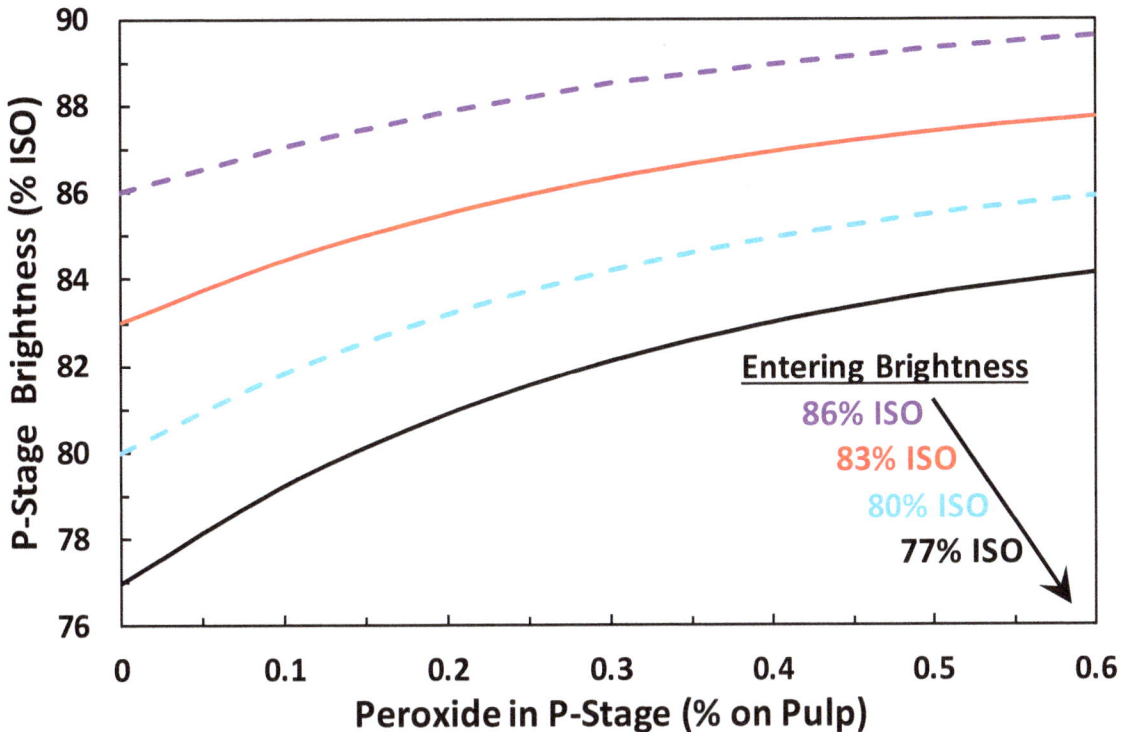

Figure 5.21. Brightening response of a United States softwood kraft pulp treated by the last peroxide stage in the $D_0(EOP)D_1P$ sequence. The entering ranged from 77% ISO to 86% ISO exiting the D_1 stage.

The brightness gains in this stage are a function of the D_1 brightness and the amount of peroxide used (Fig. 5.21). Temperatures in the peroxide stage run from 70°C to 80°C, and reaction times run up to 120 minutes. Caustic additions are on the order of 0.1 to 0.5% on pulp. Brightness gains of 3 to 6 points are typical when using 0.5% peroxide on pulp.

VII.2. Ozone Delignification

Relatively few mills in North America use ozone in their bleach plants. As of the late-2010s, there were only two mills employing ozone combined with chlorine dioxide delignification (*i.e.*, Z/D). Approximately 30 mills worldwide during this time were using ozonation to treat oxygen delignified kraft pulps, primarily from eucalypt and mixed hardwood species.

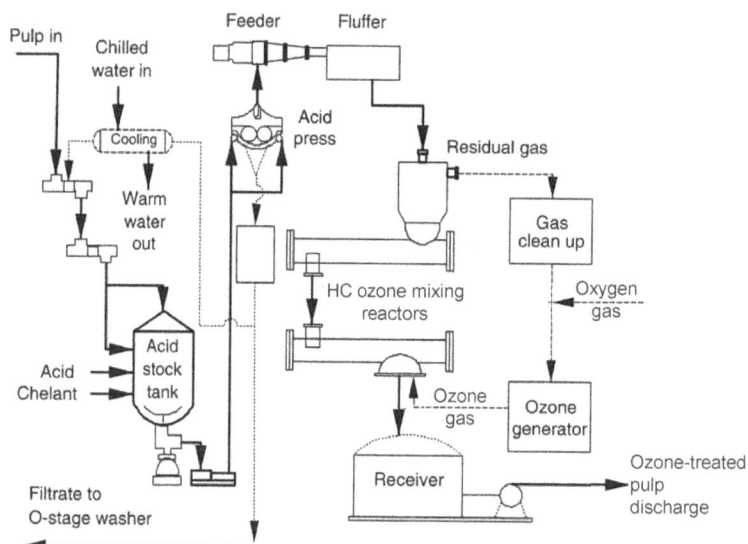

Figure 5.22. High consistency ozone reactor.

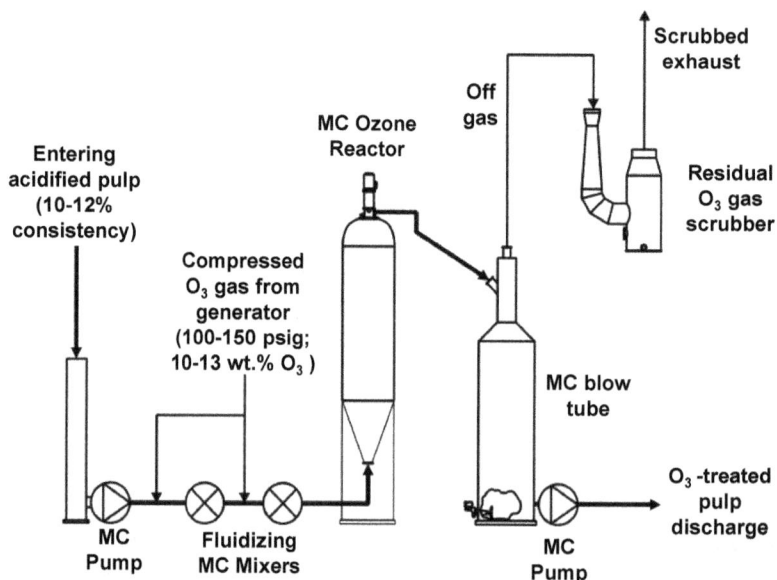

Figure 5.23. Medium consistency ozone reactor.

Ozone is a very potent oxidizing agent and is comparable to molecular chlorine. Ozone and its intermediate species, such as the hydroxyl radicals, react rapidly and indiscriminately with chemical pulps. To minimize carbohydrate attack, ozonation needs to occur at pH 2 to 4 and 40°C to 60°C. An ozone stage is performed either at medium consistency (10% to 15%) or high consistency (25% to 35%) (Figs. 5.22 and 5.23). The amount applied is limited to a maximum of 1% on pulp; its dosage is usually between 0.2% and 0.7% on pulp. Approximately 90% or higher of the ozone is consumed. Reaction times are quick, lasting from a few seconds up to 2 minutes. Figure 5.24 presents common delignification levels for oxygen delignified kraft pulps. The installation cost of an ozone delignification system in the United States processing 1000 metric tons per day ranges from 7 to 20 million US dollars (in the late 2010s).

90

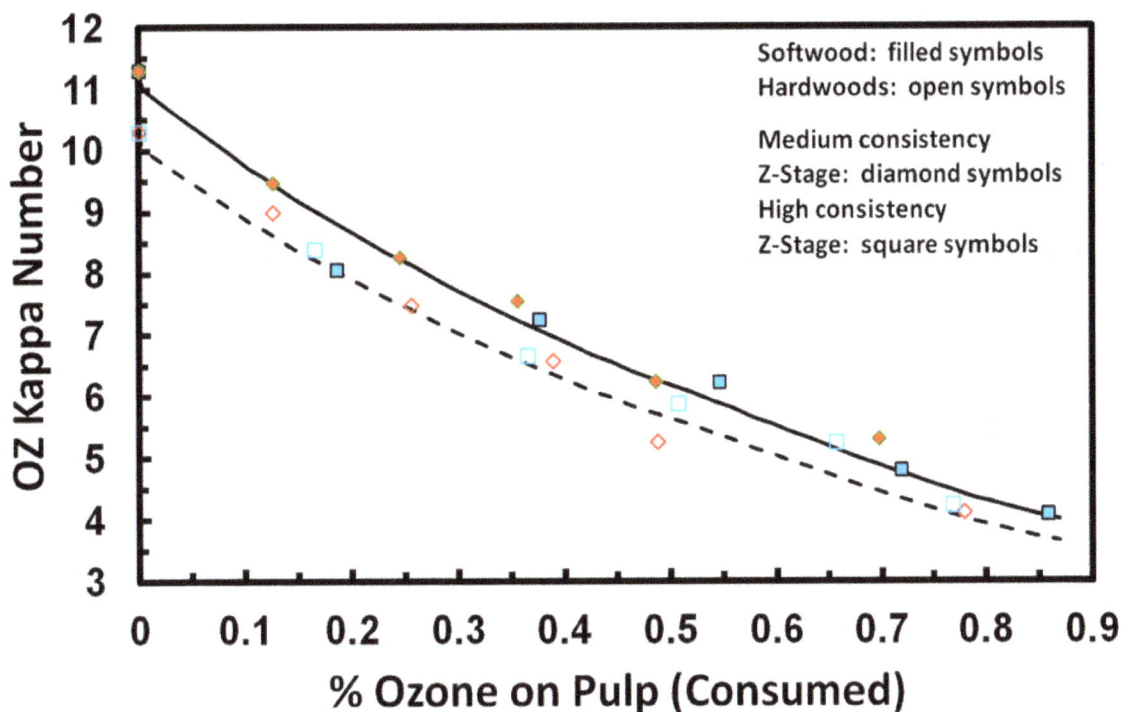

Figure 5.24. Kappa number reductions of oxygen delignified Canadian softwood and hardwood pulps after treatment with an ozone stage. Pulp viscosity decreases from *ca.* 22 mPa•s to 15 mPa•s for the softwood, and from *ca.* 15 mPa•s to 11 mPa•s for the hardwood for a 0.5% ozone charge (consumed) on pulp.

VII.3. Hot Acid Hydrolysis and Hot Chlorine Dioxide Delignification Stage

A hot acid hydrolysis (A) or a hot chlorine dioxide delignification ((D/A), D* or D_{HT}) stage is a pulp treatment. These stages are infrequently used in North American mills but are sometimes found in South American, Asian, and European mills. Their purpose is not to remove residual lignin, although these stages may reduce the stock's kappa number. Hot acid hydrolysis is used as a treatment for oxygen delignified hardwood kraft pulps, particularly eucalypt and birch species.

A hot acid hydrolysis stage removes a component known as hexenuronic acid (HexA). HexAs are formed from the uronic acid side chains of xylans during kraft pulping. These components contain a carbon double bond that consumes certain oxidants, for example molecular chlorine, hypochlorous acid, and ozone. HexAs are non-reactive towards oxygen, peroxide and chlorine dioxide itself. However, such structures react with intermediate bleaching species generated from chlorine dioxide, like hypochlorous acid. HexAs also consume permanganate and thus contribute to the measured kappa number of the pulp just like residual lignin.

Industrial conditions for a hot acid hydrolysis stage are 85°C to 100°C at a pH between 3.0 and 3.5. The stage is performed at medium consistency (10% to 13%) for 120 to 240 minutes. Up to 80% of the HexA structures are removed.

Another derivation of a hot acid hydrolysis stage is the hot chlorine dioxide delignification stage. In this configuration, the chlorine dioxide delignification stage is performed at temperatures between 85°C and 90°C instead of 45°C to 60°C. Other conditions are that the stage is conducted for 120 to 180 minutes instead of 30 to 60 minutes at medium consistency. Most all the chlorine dioxide reacts with residual lignin during the first 20 to 40 minutes, while during the remaining 100 to 120 minutes the acid conditions react to remove HexA structures. Figure 5.25 illustrates the HexA removal with an oxygen delignified eucalypt pulp.

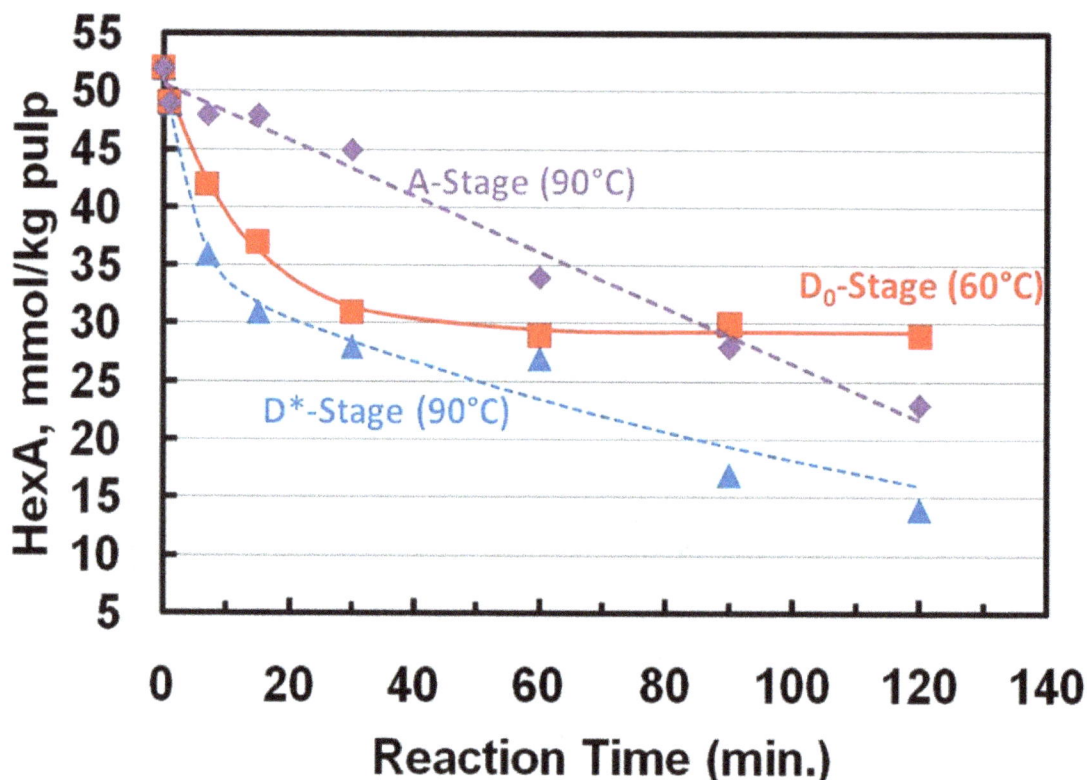

Figure 5.25. Hexenuronic acid removal for an oxygen delignified eucalypt pulp treated by a hot acid hydrolysis (A; 90°C), standard chlorine dioxide delignification (D_0; 60°C), and hot chlorine dioxide delignification (D^*, D_{HT}, or D_0/A; 90°C).

A hot acid hydrolysis or hot chlorine dioxide delignification stage can provide savings in total chlorine dioxide usage in ECF sequences that brighten to 88% to 90% ISO (Fig. 5.26). These savings can be appreciable if the HexA are above 30 mmol/kg pulp, such as those found in eucalypt, birch, mixed hardwood pulps.

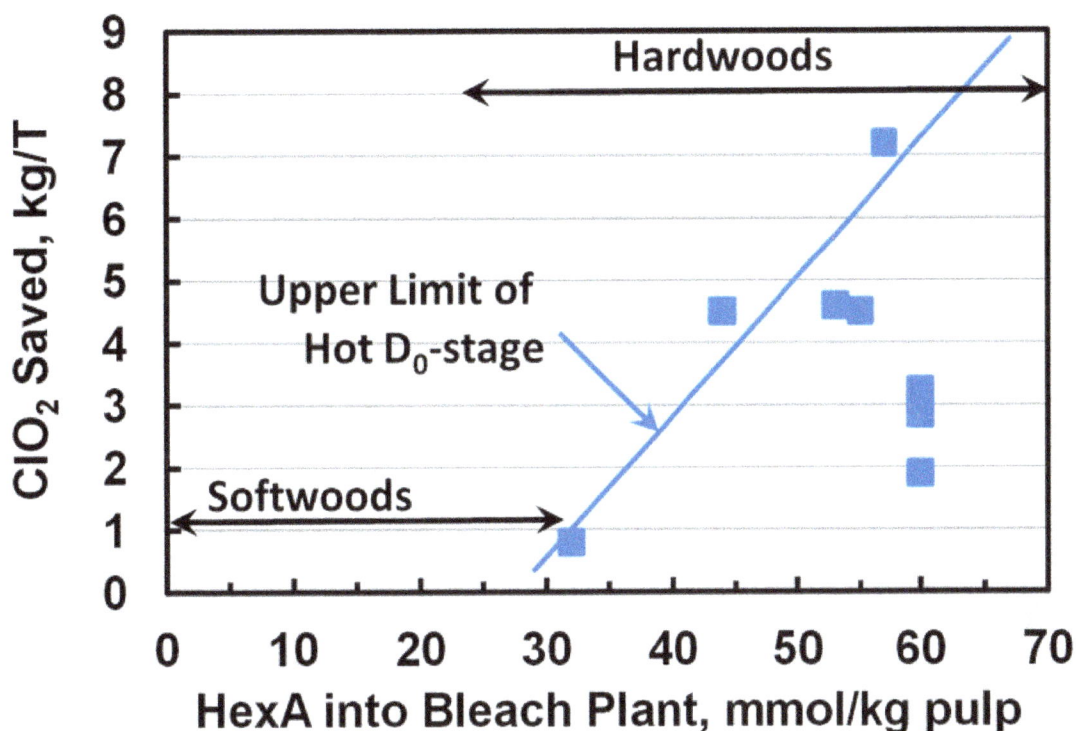

Figure 5.26. Amount of chlorine dioxide saved in ECF bleaching by using a hot acidic hydrolysis or hot chlorine dioxide delignification stage when bleaching to 88% to 90% ISO.

VII.4. Xylanase

In the early- to mid-2010s, xylanase treatment of kraft pulps was around 10 to 15 bleach plants in the United States and Canada. The endo-xylanases used are jelly-roll in shape and composed of slightly over 200 amino acids connected in series. A common addition is at the medium consistency pump to the brownstock or post-oxygen delignification storage tank. Depending on the enzyme supplied, the pulp has its pH adjusted with acid or carbon dioxide to between 6 and 8. The pulp slurry is allowed to react for 30 to 60 minutes at temperatures between 40°C and 80°C. The xylanase added is controlled to release between 3 and 6 mg xylose/g pulp.

Most xylanase used in the 2000s was derived from fungi and functioned under slightly acidic to slightly alkaline conditions. In the mid- to late-2010s, some bacteria-derived xylanases are noted to be more compatible with the alkaline conditions and temperatures typical of pulp storage. The amount of chlorine dioxide used to delignify and brighten kraft pulps can be reduced by 10% to 20% by such treatments. Using xylanase can cause 0 to 2 kappa number reductions. Where there is a small kappa number drop, the decrease is partially attributable to HexA removal. Xylanase treatment increases the chemical and biological oxygen demands in the bleaching effluents and results in pulp yield losses of up to 0.5% on pulp.

93

VII. Bleaching of Shives and Knots

Up to this point, the discussion has focused on delignification and brightening of pulp fibers. Invariably, pulp shives, knots, bark, and particles come into the bleach plant along with the pulp. These components, unlike fibers, are not bleached to the same extent. This results in dark specks in the final product.

It is preferable to remove such contaminants mechanically from the pulp by screening and cleaning before bleaching. When mechanical separations fail, the only option left is to bleach out these contaminants. Of all the bleaching contaminants, shives are the most common. Shives are defined as small splinters of undercooked wood. They are formed during kraft pulping when thick chips are not completely penetrated by cooking liquor.

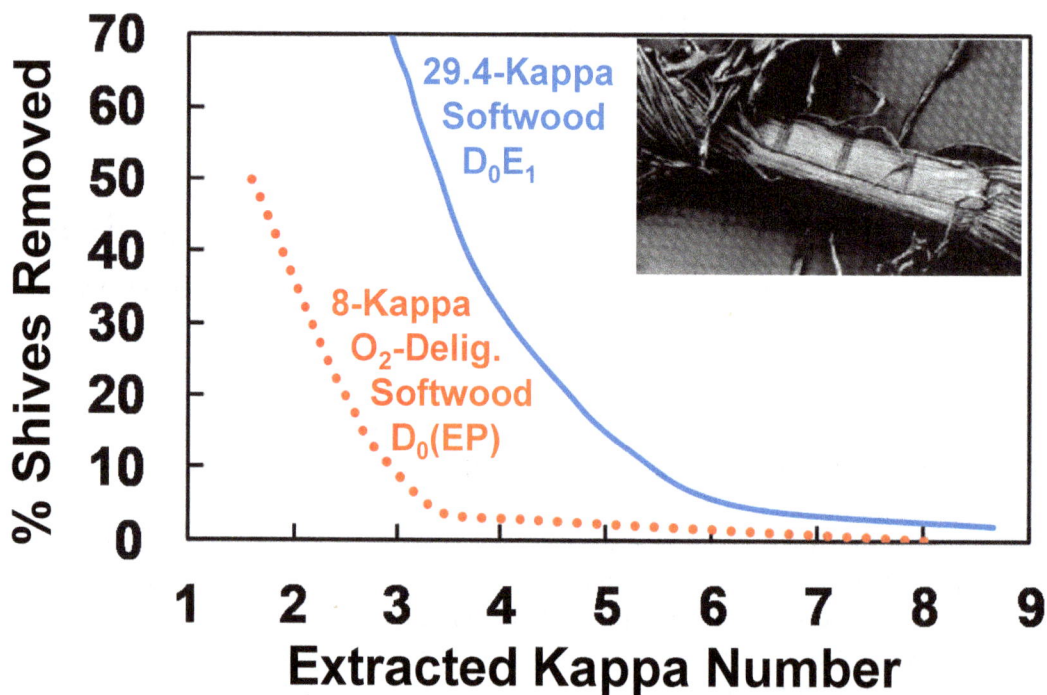

Figure 5.27. The removal of pulp shives from brownstock and oxygen delignified softwood kraft pulps as a function of the extracted kappa number.

The chlorine dioxide delignification stage can remove some shives as it delignifies the pulp. However, delignification occurs much more rapidly than shive reduction (Fig. 5.27). Shive removal does not occur to any appreciable extent until the pulp fibers have been extensively delignified to low residual lignin levels. This is because shive removal is a diffusion control process. Shives are more effectively removed in the chlorine dioxide brightening stages, where there are longer chlorine dioxide reaction times and the pulp has much lower lignin content. However, the chlorine dioxide brightening stage must be operated at a lower pH value of 3 than the optimum value around 4.

Combinations of chlorine dioxide stages are very effective at reducing shives (Fig. 5.28). Peroxide can remove shives, but not as effective as chlorine dioxide. Ozone is the least effective, primarily because it reacts so rapidly with lignin and cannot penetrate the fiber bundles. Somewhat similar trends are noted for knots (Fig. 5.29), which are partially cooked wood chips from the junctions of branches to the main trunk.

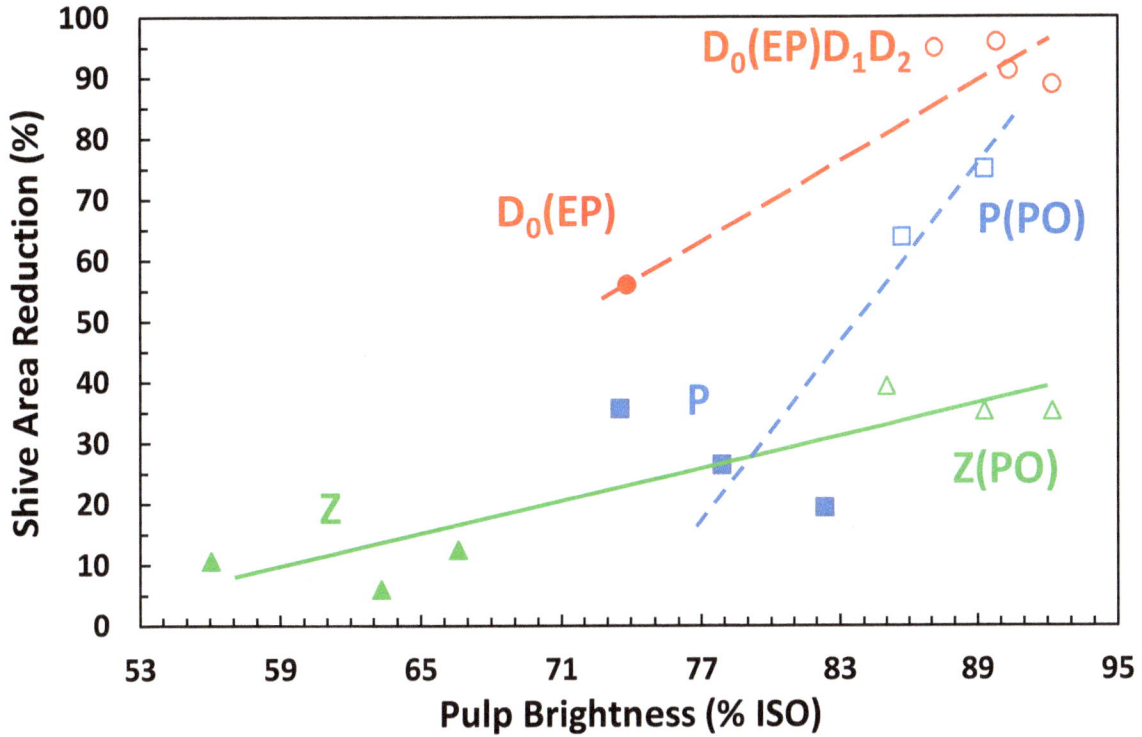

Figure 5.28. Shive area reduction as a function of bleach brightness with chlorine dioxide (D), peroxide (P), pressurized peroxide (PO), and ozone (Z) stages for an oxygen-delignified softwood kraft pulp.

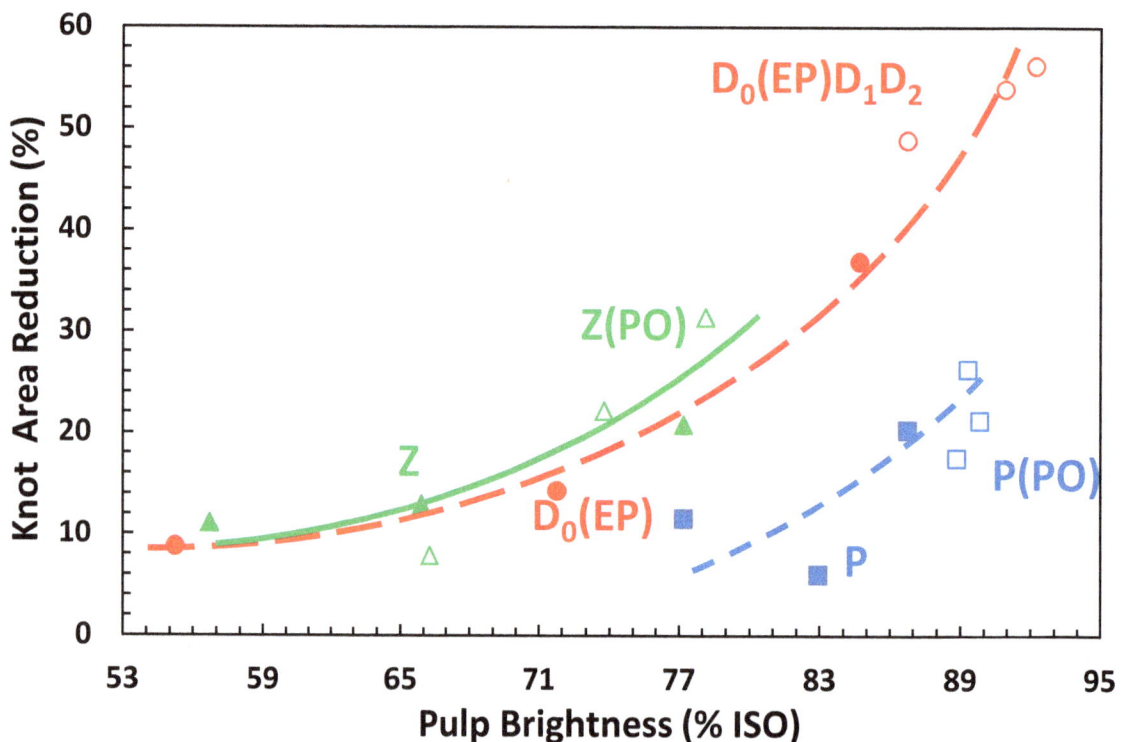

Figure 5.29. Knot area reduction as a function of bleach brightness with chlorine dioxide (D), peroxide (P), pressurized peroxide (PO), and ozone (Z) stages for an oxygen-delignified softwood kraft pulp.

Non-wood contaminants, such as fly ash, grease, oil, plastic, and rust, cannot be removed by bleaching oxidants. Most of these non-wood impurities can be eliminated by screening and screening.

Homework

1. What is the typical delignification range with an oxygen stage processing softwoods? Likewise, what is it with mixed hardwoods?
2. What is the kappa factor charge in chlorine dioxide delignification? What is another name is it known by?
3. What are common kappa factor charges used in bleaching softwood kraft pulps in the US? What is the common pH used in chlorine dioxide delignification stage?
4. An unbleached softwood kraft pulp (24 kappa number, 27.mPa•s viscosity, and 23% ISO brightness) is being bleached by the $D_0(EOP)D_1ED_2$ sequence to 89% ISO brightness. The mill is processing 800 short tons/day. The D_0 stage is using 22.9 pounds/minute of chlorine dioxide for 45 minutes, 122°F (50°C) and 3 pH (terminal). What is the kappa factor being employed in the D_0?
5. Why do we use caustic (alkali) extraction after chlorine dioxide delignification stage? Why do we reinforce this stage with oxygen and/or peroxide?
6. The D_0 pulp from question 4 is being treated by an (EOP) stage operating for 60 minutes using 5.55 pounds/minute of oxygen, 2.22 pounds/minute hydrogen peroxide, 22.0 pounds/minute caustic soda, 158°F (70°C) and 10.7 pH. Afterwards, the extracted pulp has a brightness of 45.0% ISO and kappa number of 4.2. What percentage of oxygen is

being used in the (EOP)? What percentage of hydrogen peroxide is being used? What percentage of caustic soda is being used? What is the caustic multiple being used? [Note that the caustic multiple is equal to percentage of caustic being used divided by percentage of active chlorine being used in the D_0.]

7. What is commonly used to brighten kraft pulps after bleaching delignification in the United States? What is the common pH that is used for these stages?

8. The pulp from question 6 is being bleached in a D_1 stage using 13.3 pounds/minute of chlorine dioxide for 180 minutes at a terminal pH of 4 and 158°F (70°C). The pulp had a exit brightness of 83.8% ISO. What is the percentage of chlorine dioxide being used in D_1? What is the chlorine dioxide multiple being used? [Note that the chlorine dioxide multiple is equal to percentage of chlorine dioxide being used divided by extracted kappa number of the pulp.]

9. What is hexenuronic acid (HexA)? What is it impact during bleaching? What bleaching agents directly react with HexA? What agents do not react with HexA?

10. Which bleaching agent is the most efficient at removing pulp shives? Which position in the bleach sequence is the best for removal? What is the optimum terminal pH?

References

The Bleaching of Pulp, 5th Edition, P.W. Hart and A.W. Rudie, Editors; TAPPI Press, Atlanta (2012).

Pulp Bleaching: Principles and Practice, C.W. Dence and D.W. Reeves, Editors; TAPPI Press, Atlanta (1996).

Pulp Bleaching Today, H.U. Suess; Walter de Gruyter GmbH & Co. KG, New York (2010).

2023 TAPPI Bleach Plant Operations Workshop, B.N. Brogdon, Lead Instructor; TAPPI Press, Atlanta (2023).

Chapter 6: Production of Bleaching Chemicals

I. Introduction

The earlier chapter covered the bleaching of kraft softwood and hardwood kraft pulps with various delignification and brightening agents. This chapter will discuss how the various oxidants and auxiliary chemicals are manufactured or purchased.

II. Production of Chlorine Dioxide

Chlorine dioxide is produced on site by the reduction of sodium chlorate in a highly acidic medium (above 6N H^+). Most mills use either an ERCO Worldwide R8 generator or a Nuryon SVP-LITE generator. These generators use methanol as the reducing agent and operate under subatmospheric (*i.e.*, vacuum) conditions. Other chlorine dioxide generators, such as SVP-HP, use hydrogen peroxide instead of methanol as the reducing agent.

Chlorine dioxide can be made by the acidification of sodium chlorite. This method is not economical for producing chlorine dioxide for kraft pulp bleaching. Instead, this process is most often used in laboratories to generate small amounts of chlorine dioxide because of its simplicity.

The stoichiometric conversion of sodium chlorate to chlorine dioxide in the R8 and SVP generators is given as:

$$9NaClO_3 + 6H_2SO_4 + 2CH_3OH = 9ClO_2 + 3Na_3H(SO_4)_2 + 0.5CO_2 + 1.5HCOOH + 7H_2O$$

    sodium   sodium   formic
    chlorate   sesquisulfate   acid

Typically, the reaction does not go to completion, and the chlorine dioxide yields are around 80 to 98%. A byproduct of the process is the double salt of sodium sulfate and sodium bisulfate, known as sodium sesquisulfate. Sodium sesquisulfate can be crystalized to separate it from the spent generator liquor by filtration. Adding some water to this double salt can cause it to break down or metathesis to precipitate sodium sulfate and release sulfuric acid:

$$2Na_3H(SO_4)_2\ (aq) = 3Na_2SO_4\ (s) + H_2SO_4\ (aq)$$

Or the sodium sesquisulfate can be neutralized with caustic to precipitate only sodium sulfate:

$$Na_3H(SO_4)_2\ (aq) + NaOH\ (aq) = 2Na_2SO_4\ (s) + H_2O$$

This sodium sulfate from either process is used to compensate for some of the sulfur losses in brownstock washing.

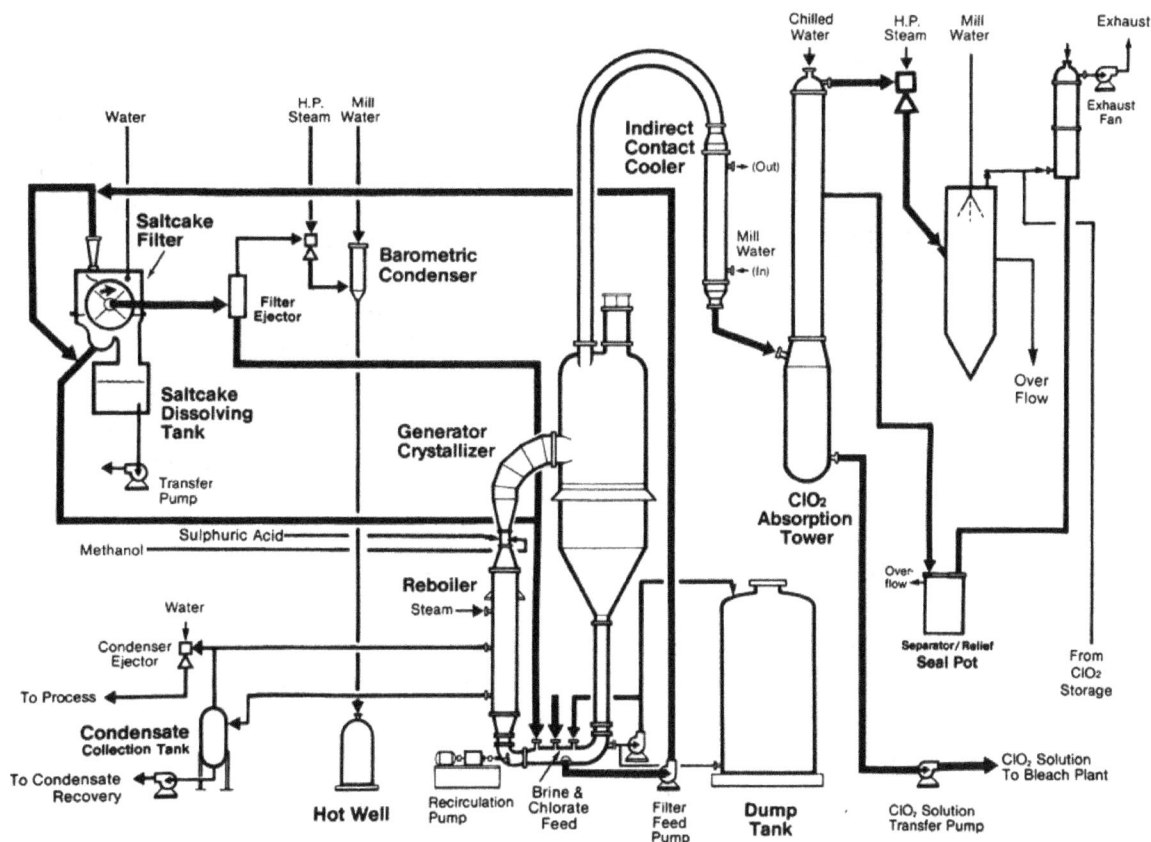

Figure 6.1. Example of a chlorine dioxide generator (R8 or SVP-LITE process) at the pulp mill. [Reproduced from the *1989 TAPPI Bleach Plant Workshop*, TAPPI Press, Norcross, GA, p. 155.]

Table 6.1. Raw materials and utilities used in modern chlorine dioxide generators, as well as credited sodium sulfate (salt cake). (SVP is single vessel process, and SCW is salt cake wash.)

Raw materials	Units	Generator Type	
		R8/SVP-LITE	R10/SVP-SCW
Sodium chlorate	kg/kg ClO_2	1.75	1.75
Methanol	kg/kg ClO_2	0.16	0.16
Sulfuric acid	kg/kg ClO_2	1.08	1.08
Salt Cake			
Sesquisulfate produced	kg/kg ClO_2	1.44	0.00
Caustic soda (neutralization)	kg/kg ClO_2	0.22	0.02
Sodium sulfate (salt cake) credit	kg/kg ClO_2	1.61	1.10
Utilities			
0.4 MPa steam	kg/kg ClO_2	2.72	0.45
1.0 MPa steam	kg/kg ClO_2	4.53	5.10
Electricity	kW•hr/kg ClO_2	0.65	0.62
Chilled water	L/kg ClO_2	37.2	37.2

Figure 6.1 is a diagram of the R8 or SVP-LITE process. Sodium chlorate is added, heated with steam at the reboiler and mixed with sulfuric acid and methanol. This mixture is sent to the generator, where chlorine dioxide gas is formed. The gas is sent to a cooler, and then to an

absorption tower where the chlorine dioxide gas is dissolved into chilled water to form a solution that is between 6 and 12 g/L. A stream of the spent brine from the generator is sent to the salt cake filter to recover the solid sodium sesquisulfate. This recovered solid is redissolved in water in the dissolving tank to make sodium sulfate. The separated spent liquid at the salt cake filter is recycled to the reboiler. A steam ejector is used to create a vacuum for the generator and absorption tower, as well as for the salt cake filter. Newer generator designs, such as R10 or SVP-SCW, are variations of R8 or SVP-LITE regarding the salt cake process.

Table 6.1 provides the general raw materials and utilities consumed by the generator per kilogram of chlorine dioxide produced. The table also affords information about the sodium sulfate made, which is sent to chemical recovery and used as a credit. The approximate cost of chlorine dioxide in the early 2020s is around 1.35 to 1.50 USD/kg. The largest cost contributor, approximately 80%, is the sodium chlorate consumed. It takes 1.75 kg of sodium chlorate for each kilogram of chlorine dioxide produced. The sodium chlorate is supplied to the mill as a 40% to 45% concentrated solution, or as a solid crystalline product. The capital cost for a chlorine dioxide generator runs from 18 to 21 million US dollars for a mill producing 1000 metric tons/day pulp.

Chlorine dioxide is supplied from the bleach plant is dissolved in water, usually around 10 g/L. It is an unstable reagent in its gas phase. Chlorine dioxide gas can undergo spontaneous decomposition reactions called puffs (at less than 1 m/s) if its concentration reaches a partial pressure of 100 to 120 mm Hg (0.133 to 0.160 bar). Therefore, chlorine dioxide generators operate under subatmospheric conditions with steam (or air) to sweep out and dilute the formed gas. Chilled chlorine dioxide solutions (*ca.* 5°C) are stable for months if kept in the dark.

III. Oxygen Production

Oxygen gas is produced on site at the bleach plant by its concentration from nascent air. There are two commercial operations used: cryogenic separation and adsorption. Atmospheric air is purified to remove residual moisture and carbon dioxide. This is done by filtering the air to remove particulate matter. Then, the modified air is compressed and refrigerated multiple times to remove the traces of moisture and carbon dioxide. The purified air is subjected to multiple treatments of hot and cold heat exchangers connected in series to separate the air into oxygen and nitrogen components. After this process, the chilled air is subjected to distillation, which produces the purified oxygen gas.

The second method of generating oxygen gas is by adsorption. This is a relatively simple process able to produce 95% oxygen. The commercial process uses a cyclic sequence where the nitrogen from the incoming stream is adsorbed onto molecular sieves to result in an enriched oxygen, and then the molecular sieves are de-adsorbed to release the waste nitrogen. This process can be conducted by pressure swing adsorption (PSA) where adsorption is done under pressure and the regenerative de-adsorption is performed at atmospheric conditions. Alternatively, the process can be done by vacuum swing adsorption (VSA) where the adsorption is done at near atmospheric pressure, and the de-adsorption is conducted at vacuum pressure. Yet a third method, vacuum pressure swing adsorption (VPSA), is where the adsorption occurs under pressurized conditions and the de-adsorption is conducted under a vacuum.

If the oxygen gas is not generated on-site by cryogenic or adsorption, it can be purchased separately. The gas is typically bought from a chemical supplier and is provided as a refrigerated liquid. The capital cost for an oxygen generator is about 6 to 10 million US dollars for a mill that produces1000 metric tons/day pulp.

IV. Ozone Production

Ozone gas is manufactured at the bleach plant from oxygen. Because of its high reactivity, ozone cannot be stored and transported. It is produced through an electrical corona discharge process. Purified oxygen gas is passed through a corona discharger operating at 400 to 500 Hz. The result is ozone gas that is 10% to 12% by weight. The electricity needed is about 10 kW•hr/kg ozone produced. The capital cost for an ozone generator is about 21 million US dollars for a mill manufacturing 1000 metric tons/day pulp.

V. Hydrogen Peroxide Solution

Hydrogen peroxide is supplied to the bleach plant as concentrated solutions of 50% by weight. These concentrated solutions can undergo catalytic decomposition reactions. The decomposition can become uncontrollable, generating significant amounts of heat and gases. The amount of explosive energy released is nearly equivalent to that of dynamite. Over the past twenty-five years, there have been several catastrophic bleach plant instances of sudden pressure releases with 50% hydrogen peroxide. It is recommended that 50% solutions from bulk storage should be diluted to 5% to 10% by weight and stored in intermediate dilution tanks just prior to their use. The benefits of using more concentrated hydrogen peroxide are small to negligible when less than 1% peroxide is used, which is most common in ECF bleach sequences.

VII. Caustic Soda (Sodium Hydroxide) Solution

Bulk sodium hydroxide or caustic soda solutions are purchased by the mill at 50% by weight. These solutions are often diluted down to 5% up to 25% by weight just prior to their use. There are three different grades available for use. The mercury-cell grade has the highest purity and has the lowest iron ion levels (< 1 part per million (ppm)). The membrane-cell grade is the midline grade and contains between 1 and 5 ppm of iron. The least expensive grade of caustic soda is diaphragm grade. It has 5 to 10 ppm of iron. If caustic is used with hydrogen peroxide, it is most preferable to use the mercury- or membrane-cell grade to prevent catalytic peroxide decomposition.

VII. Sulfuric Acid Solution

Sulfuric acid, like caustic soda, is frequently used in bleach plants for pH adjustments, and in the generation of chlorine dioxide. It is supplied to the mill as a concentrate at 93% to 98% strength. Sulfuric acid is produced from sulfur dioxide and oxygen *via* a catalytic oxidation process to form sulfur trioxide, which is subsequently adsorbed into water.

Homework

1. What are the most common chlorine dioxide generators used in the United States?
2. How much sodium chlorate ($NaClO_3$) is needed to produce 1 lb (or kg) of chlorine dioxide?
3. What are the byproducts generated by a chlorine dioxide generator? What is the byproduct used for in kraft pulping?
4. What are the two methods generally used to produce oxygen gas used in chemical pulp bleaching?
5. How is ozone produced? At what concentration range is the ozone produced at with this process?
6. What are the purchased chemicals used in kraft pulp bleaching? What are their chemical concentrations at purchase?

References

The Bleaching of Pulp, 5th Edition, P.W. Hart and A.W. Rudie, Editors; TAPPI Press, Atlanta (2012).

Pulp Bleaching: Principles and Practice, C.W. Dence and D.W. Reeves, Editors; TAPPI Press, Atlanta (1996).

Pulp Bleaching Today, H.U. Suess; Walter de Gruyter GmbH & Co. KG, New York (2010).

2023 TAPPI Bleach Plant Operations Workshop, B.N. Brogdon, Lead Instructor; TAPPI Press, Atlanta (2023).

1989 TAPPI Bleach Plant Operation Short Course; TAPPI Press, Norcross (1989).

Chapter 7: Bleach Plant Emissions

I. Introduction

Bleaching requires the pulp to be washed to remove dissolved organics and spent chemicals from each stage. Unlike spent black liquor from kraft pulping, this aqueous waste is typically not sent back to chemical recovery. This is because the effluent contains chlorinated organic and inorganic compounds. These compounds are incompatible in the kraft recovery boiler. Instead, this effluent is processed through wastewater treatment to remediate it before it is released back into the environment.

Emissions formed during bleaching are regulated by United States authorities to ensure that a pulp mill complies with federal and local codes. Parameters that are used to characterize effluents include biological oxygen demand (BOD), chemical oxygen demand (COD), color, adsorbable organic halide (AOX), and chronic and acute toxicity. The following section covers some specifics.

II. Biological Oxygen Demand (BOD) and Chemical Oxygen Demand (COD)

Bleaching effluents contain dissolved organic compounds. Some of these compounds can be degraded and metabolized by microorganisms (or bacteria), either during wastewater treatment or in the receiving streams. With these biological processes, the dissolved oxygen is consumed by the microorganisms. Depletion of dissolved oxygen can adversely affect larger aquatic organisms found in the receiving streams, such as fish and aquatic plants. These components are broadly measured using BOD and COD tests.

The BOD test provides an estimate of the effluent's propensity to deplete dissolved oxygen in the receiving waters. It measures how much of the dissolved organics is biodegradable. BOD can be decreased by: lowering pulp yield losses (or bleach shrinkage), or lowering the pulp's kappa number entering the chlorine dioxide delignification stage. This is done either through oxygen delignification and/or modifications to kraft pulping processes. The BOD in treated effluents is nearly eliminated once the wastewater has undergone biological treatment. Reported BOD levels are given as mg O_2/L or kg/t pulp. The test can be performed for 5 days (BOD_5) or 7 days (BOD_7).

Another test that measures dissolved organics in effluents is COD. This test differs from BOD in that the test is much quicker to perform, taking a few hours to complete *versus* 5 to 7 days. The COD test is a more complete measurement of the dissolved organics. This test employs a very strong chemical oxidant, such as potassium dichromate, to oxidize the organic materials under standardized laboratory conditions. Measured COD levels are higher than those detected by BOD. That is because more dissolved organics can be oxidized by potassium dichromate than can be oxidized and metabolized by bacteria. COD, in combination with BOD, can quantify portions of dissolved organic materials that are recalcitrant, or resistant, towards biodegradation in

103

primary and secondary effluent treatment. Measured COD in effluents is proportional to the un-bleached or oxygen delignification kappa number entering the bleach plant, as shown in Figure 7.1 (*ca.* 2 kg O_2 COD/kappa unit). The units of measurement are mg O_2/L or kg/t pulp.

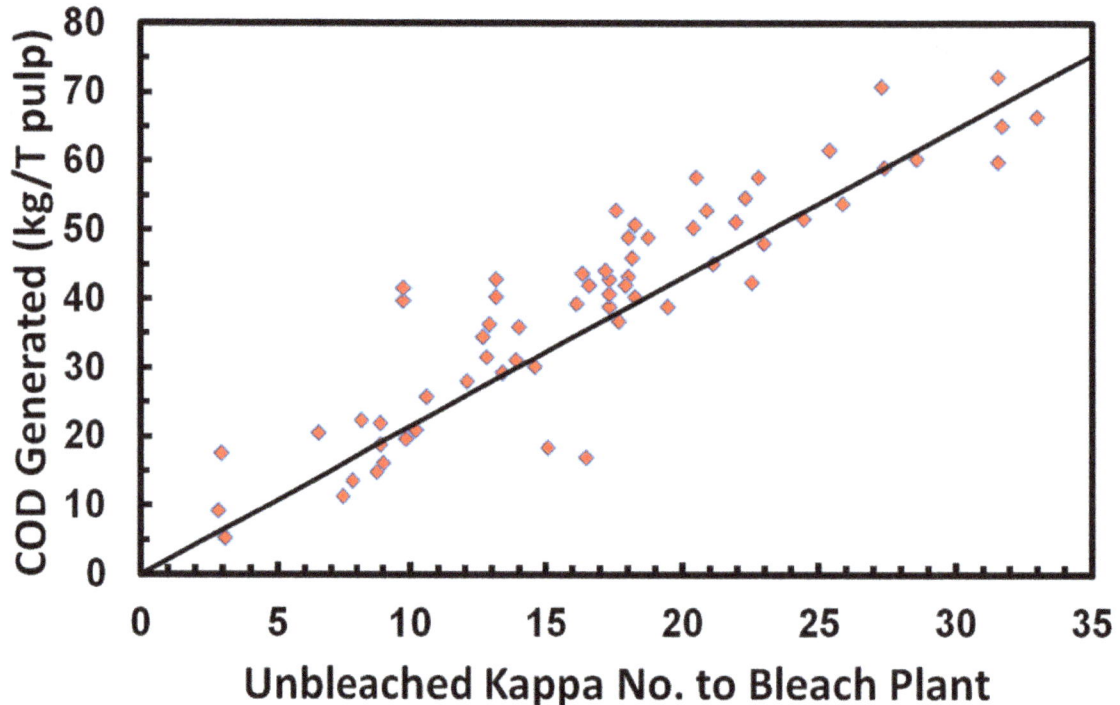

Figure 7.1. COD generated during bleaching as a function of the entering kappa number into the bleach plant.

III. Total Suspended Solids (TSS)

Total suspended solids (TSS) are solid materials that are dispersed or slurried in the wastewater. It primarily consists of fiber fragments lost during processing. A sample of effluent is filtered, and the quantity of dry solids is given as TSS expressed as mg/L or as kg/t pulp.

IV. Color

Color is an important characteristic of bleaching effluent. The main components that contribute to this physical descriptor are lignin and its degradation products. Color imparts an aesthetically undesirable appearance to receiving water. More importantly, it has a negative impact on aquatic life by blocking sunlight that is needed for photosynthesis in plants. Color is quantified by measuring the waste sample against color standards, such as cobalt chloroplatinate. Most color found in combined bleach plant effluents is attributed to dissolved organic materials from the first extraction stage. This parameter is not federally regulated in the United States, but it is observed in select regional jurisdictions. Color is unaffected by secondary wastewater treatment. Like COD, color is proportional to the brownstock or oxygen-delignified kappa number of the pulp entering the bleach plant. It is measured and reported as mg chloroplatinate/L or kg/t pulp.

V. Adsorbable Organic Halide (AOX)

The next effluent parameter is adsorbable organic halide content, or AOX. AOX is a parameter that measures the amount of organically-bound chlorine in the dissolved organics of the sample. The reason AOX is used is that some chlorine-containing compounds are toxic to animal life, most notably polychlorinated dioxins and highly chlorinated phenols. It should be noted that not all chlorinated-organic materials are toxic to life. So, a measured AOX level may or may not specifically correlate to an effluent's toxicity.

AOX is measured by a given procedure. Granulated activated carbon is mixed with the effluent sample, and the chlorinated organic components are adsorbed onto the solid. Afterwards, the activated carbon with its adsorbed organics is filtered to collect the solids. These solids are burned with oxygen using a combustion micro-coulometer instrument, where the mineralized HCl formed is measured *via* titration. Measurement values are given in micrograms/L or kg/t pulp.

VI. Chloroform and Total Organic Carbon (TOC)

Chloroform generation in current ECF bleach plants is low compared to the 1990s or earlier, when the use of molecular chlorine and sodium hypochlorite were common. However, bleach plant effluents still have to be evaluated for chloroform levels. Values are reported in micrograms/L. Another common test performed on effluents is total organic carbon (TOC). It is measured by instrumental or chemical test methods. Specific relationships between TOC and BOD levels for effluents exist but must be determined individually for each bleaching effluent. It should be noted that all carbon compounds detected by TOC are oxidized by microorganisms nor are they chemically oxidized. Chloroform levels are measured in mg/L or g/t pulp.

VII. Individual Compounds

The final group of effluent tests is for individual compounds. The first of these is for the detection of polychlorinated organic compounds such as phenols, guaiacols, catechols, and vanillins substituted with three or more chlorine atoms. These chlorinated phenolics are detected at the microgram/L level. The second group detects minute traces of polychlorinated dioxins and furans, such as 2,3,7,8-tetrachlorodibenzo-*p*-dioxin (2,3,7,8-TCDD) and 2,3,7,8-tetrachloro-dibenzofuran (2,3,7,8-TCDF). These chlorinated dioxin and furan compounds are reported as picograms/L.

VIII. Sum versus Effect Parameters

The wastewater tests can be categorized into two broad classes. Sum parameters afford a synopsis measure for a group of compounds in a sample, without distinguishing individual components. Such tests asses the overall pollution levels into a single value. Effect parameters, on the other hand, directly measure the biological effects of effluents on organisms. Tests that fall into this category give information about the toxicity, genotoxicity, or other ill impacts of a contaminant. Effluent tests that are classified as sum parameters are BOD, COD, TSS, and AOX. Tests that measure single-species acute and short-term toxicity fall into the effect parameters category.

IX. Toxic Substances

Toxic substances found in kraft bleaching effluents are in much, much lower quantities than oxygen-consuming substances, as denoted by BOD and COD. Such compounds include resin acids, fatty acids, and terpenes found in wood and kraft pulps. These extractives are an organic and natural pesticide that a tree produces to ward-off microorganism and wildlife attacks. Toxic substances also include the polychlorinated phenolic compounds, dibenzo-*p*-dioxins and dibenzofurans formed during bleaching. An effluent's toxicity can be acute, which means the death of the specimen within a short time, or chronic where the effects can be more subtitle and serious. Such toxicity measurements are performed with specific bioassays, such as with water daphnia, fathead minnows and sea urchins.

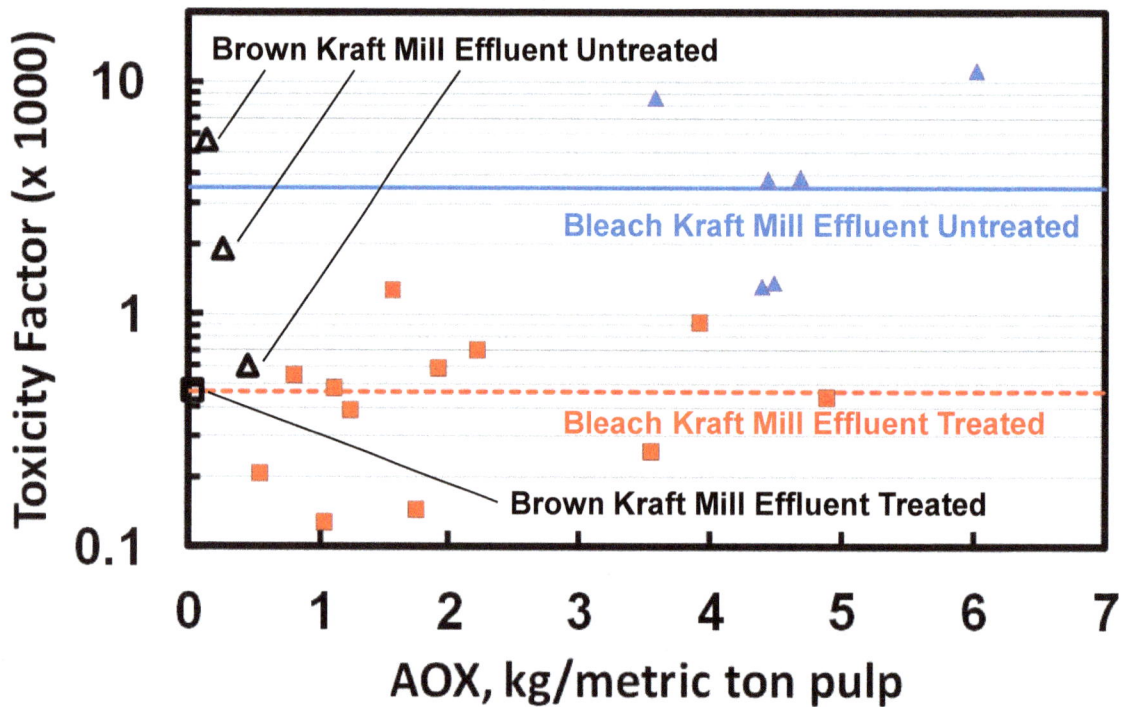

Figure 7.2. Measured AOX from bleached and brown Canadian kraft mills before and after effluent treatment *versus* its toxicity with water daphnia reproduction.

Figure 7.2 presents effluent toxicity from Canadian kraft pulp mills as a function of AOX. The toxicity before and after biological treatment, as well as its corresponding AOX, was measured. The results showed that toxicity was indistinguishable between the mills based on measured AOX levels from 0.5 to 6.0 kg/t pulp. There was only a slight effect observed between the untreated and treated wastewater samples.

X. United States Cluster Rule Regulations

In 1998, the United States promulgated the Cluster Rule regulations. These rules integrated water and air emission limits for the pulp and paper industry. Compliance with these federal regulations went into effect in 2001. These rules set daily maximum and monthly average limits on treated effluents for selected sum and effect parameters. Tables 7.1 through 7.3 present some excerpts for treated bleach plant effluents.

Table 7.1. United States effluent guidelines for new bleach plants (2001 or newer) producing pulp and fine paper: AOX, BOD_5, COD and TSS limits.

Parameter	One-Day Maximum	Monthly Average
AOX	0.476 kg /t pulp	0.272 kg/t pulp
BOD_5	5.70 kg/t pulp	3.10 kg/t pulp
COD	Deferred	Deferred
TSS	9.10 kg/t pulp	4.80 kg/t pulp

Table 7.2. United States effluent guidelines for existing bleach plants producing market pulps: AOX, BOD_5, COD and TSS limits.

Parameter	One-Day Maximum	Monthly Average
AOX	0.951kg /t pulp	0.623 kg/t pulp
BOD_5	15.45 kg/t pulp	8.05 kg/t pulp
COD	Deferred	Deferred
TSS	30.4 kg/t pulp	16.4 kg/t pulp

Table 7.3. United States effluent guidelines for specific chlorine-containing compounds for existing and new bleach plants.

Parameter	One-Day Maximum	Monthly Average
2,3,7,8-TCDD	Non-detectable (<10 pg/L)	-
2,3,7,8-TCDF	31.9 pg/L	-
Chloroform	6.92 g/t pulp	4.14 g/t pulp
2,4,6-Trichlorophenol	Non-detectable (<2.5 mcg/L)	-
2,4,5-Trichlorophenol	Non-detectable (<2.5 mcg/L)	-
11 Other Polychlorophenols*	Non-detectable (<2.5 or 5.0 mcg/L)	-

*Includes all tri-, tetra- and pentachlorinated versions

The higher limits listed here are for older bleach plants producing market pulps. Other bleached pulp grades may have lower regulated limits. As of this date, the EPA does not regulate COD and color but reserves the right in the future to enact such limits. With most specific compounds, the regulations dictate that they be at non-detectable limits. This means that the components are just below the threshold at which the instrumentation can measure. For the most sensitive compounds, such as 2,3,7,8-TCDD, it is below 10 pg/L.

XI. General Comments Regarding Bleaching Effluents

BOD, COD, color and AOX values found in the effluents are all roughly proportional to the kappa number drop of the pulp as it progresses through the bleach plant. The yield loss or shrinkage of the pulp during bleaching is proportionate to BOD and COD levels. In ECF bleach plants, AOX is primarily generated in the chlorine dioxide delignification and subsequent caustic extraction. Approximately 10% of the chlorine atoms supplied by chlorine dioxide delignification end up as AOX; this is much higher than the 2% of the chlorine atoms provided by chlorine dioxide brightening stages. It should be noted that more than half of the AOX generated in raw bleach effluents is removed by biological treatment. The amount of AOX can be lowered by reducing the kappa number of pulp entering the bleach plant, and/or by reducing the amount of chlorine dioxide used for delignification.

XII. Air Emissions

Besides bleach plant effluent emissions, the EPA Cluster Rule regulations also cover the release of hazardous air pollutants (HAPs). This includes some gases used in bleaching, like chlorine dioxide and ozone, and some that are generated byproducts, such as molecular chlorine and volatile organics, like chloroform and methanol. These HAPs have to be captured and neutralized. Efficient bleach plant scrubbers are used to remove and/or neutralize captured HAPs to 99% removal efficiencies. Enclosures and closed vent systems are used for bleaching towers, washers, and filtrate tanks to prevent gaseous emissions.

Homework

1. What is biological oxygen demand (BOD)? What is chemical oxygen demand (COD)? What is total suspended solid (TSS)? What is color? What is adsorbable organic halide (AOX)?
2. What are the specific organic compounds that are regulated under the United States EPA regulations?
3. What are the two chemical agents used o bleach kraft pulps in the United States that are generally not allowed since 2003? Why where these agents banned?
4. What are the air emissions of concern during kraft pulp bleaching? At what levels are they required for removal under the United States EPA regulations?

References

The Bleaching of Pulp, 5th Edition, P.W. Hart and A.W. Rudie, Editors; TAPPI Press, Atlanta (2012).

Pulp Bleaching: Principles and Practice, C.W. Dence and D.W. Reeves, Editors; TAPPI Press, Atlanta (1996).

Pulp Bleaching Today, H.U. Suess; Walter de Gruyter GmbH & Co. KG, New York (2010).

2023 TAPPI Bleach Plant Operations Workshop, B.N. Brogdon, Lead Instructor; TAPPI Press, Atlanta (2023).

N

Na$_2$O, 19, 39, 40, 45, 46
Normality, 10, 14, 17, 18, 19

O

OD(EOP), 63
Oxidant Ranking, 61
Oxidation, 21, 52, 61, 75, 76, 101
Oxygen Delignification, 22, 23, 48, 58, 59, 60, 62, 63, 65, 69, 70, 71, 72, 76, 78, 93, 100, 103, 104
Ozone, 22, 48, 49, 53, 56, 57, 60, 61, 62, 63, 69, 89, 90, 91, 95, 96, 101, 102, 108

P

Paracrystalline, 51
Particle Removal Ability, 61
Pectin, 29
Peracetic Acid, 22
Permanganate Number, 52
Peroxide, 24, 48, 49, 53, 60, 61, 62, 63, 64, 76, 78, 79, 80, 86, 88, 89, 91, 95, 96, 98, 101
pH, 7, 8, 9, 11, 12, 13, 14, 15, 16, 20, 21, 24, 39, 49, 60, 73, 75, 76, 80, 83, 84, 85, 86, 88, 90, 91, 93, 94, 96, 97, 101
Phenolic, 30
Phenols, 105
Phloem, 31, 32
Pith, 32
pOH, 7, 11, 12
Polychlorinated Phenolic Compounds, 106
Polyphenols, 53
Primary Cell Wall, 27
Pulp Strength, 56

Q

Quinones, 53, 73

R

Radial Cut, 31
Rays, 32
Reactivity, 61
Reduction, 21, 23
Residual Lignin, 52, 65, 73
Resins, 30

S

S1 Layer, 27
S2 Layer, 27
S3 Layer, 27
Sapwood, 32
Sea Urchins, 106
Selectivity, 60, 61, 71
Sodium Chlorate, 24, 98, 100, 102
Sodium Chlorite, 98
Sodium Hypochlorite, 48, 60, 61, 105
Sodium Hypochlorite, 62
Sodium Sesquisulfate, 98, 100
Softwood, 32, 50, 51, 73, 76, 80
Solution, 5, 7, 8, 9, 10, 11, 15, 16, 17, 18, 19, 20, 21, 22, 24, 25, 35, 45, 46, 47, 52, 55, 69, 100
Strength, 33, 55, 56
Strong Acids/Bases, 7
Sulfidity, 35, 39, 40, 45, 46
Sulfuric Acid, 10, 11, 18, 52, 60, 83, 98, 99
SVP-LITE Process, 99
Syringyl, 51

T

Tangential Cut, 31
Tensile, 55
Terpenes, 30
Titration, 16, 17, 18, 20, 22, 24
Total Alkali, 16, 38
Total Suspended Solids, 104

V

Vanillins, 105
Viscosity, 55

W

Water Daphnia, 106
Waxes, 30
Weak Base, 11, 13
Whiteness, 53

X

Xylanase, 93
Xylanase, 48, 62, 93

Y

Yield Vs. Kappa Curve, 41

www.ingramcontent.com/pod-product-compliance
Lightning Source LLC
Chambersburg PA
CBHW051348200326

41521CB00014B/2519